Amr Mohamed

Utility-based Resource and QoS Optimization in Packet Networks

Amr Mohamed

Utility-based Resource and QoS Optimization in Packet Networks

link and network level optimization

VDM Verlag Dr. Müller

Impressum/Imprint (nur für Deutschland/ only for Germany)

Bibliografische Information der Deutschen Nationalbibliothek: Die Deutsche Nationalbibliothek verzeichnet diese Publikation in der Deutschen Nationalbibliografie; detaillierte bibliografische Daten sind im Internet über http://dnb.d-nb.de abrufbar.
Alle in diesem Buch genannten Marken und Produktnamen unterliegen warenzeichen-, marken- oder patentrechtlichem Schutz bzw. sind Warenzeichen oder eingetragene Warenzeichen der jeweiligen Inhaber. Die Wiedergabe von Marken, Produktnamen, Gebrauchsnamen, Handelsnamen, Warenbezeichnungen u.s.w. in diesem Werk berechtigt auch ohne besondere Kennzeichnung nicht zu der Annahme, dass solche Namen im Sinne der Warenzeichen- und Markenschutzgesetzgebung als frei zu betrachten wären und daher von jedermann benutzt werden dürften.

Coverbild: www.purestockx.com

Verlag: VDM Verlag Dr. Müller Aktiengesellschaft & Co. KG
Dudweiler Landstr. 125 a, 66123 Saarbrücken, Deutschland
Telefon +49 681 9100-698, Telefax +49 681 9100-988, Email: info@vdm-verlag.de
Zugl.: Vancouver, University of British Columbia, Diss., 2008

Herstellung in Deutschland:
Schaltungsdienst Lange o.H.G., Zehrensdorfer Str. 11, D-12277 Berlin
Books on Demand GmbH, Gutenbergring 53, D-22848 Norderstedt
Reha GmbH, Dudweiler Landstr. 99, D- 66123 Saarbrücken
ISBN: 978-3-639-04446-1

Imprint (only for USA, GB)

Bibliographic information published by the Deutsche Nationalbibliothek: The Deutsche Nationalbibliothek lists this publication in the Deutsche Nationalbibliografie; detailed bibliographic data are available in the Internet at http://dnb.d-nb.de.
Any brand names and product names mentioned in this book are subject to trademark, brand or patent protection and are trademarks or registered trademarks of their respective holders. The use of brand names, product names, common names, trade names, product descriptions etc. even without
a particular marking in this works is in no way to be construed to mean that such names may be regarded as unrestricted in respect of trademark and brand protection legislation and could thus be used by anyone.

Cover image: www.purestockx.com

Publisher:
VDM Verlag Dr. Müller Aktiengesellschaft & Co. KG
Dudweiler Landstr. 125 a, 66123 Saarbrücken, Germany
Phone +49 681 9100-698, Fax +49 681 9100-988, Email: info@vdm-verlag.de

Produced in USA and UK by:
Lightning Source Inc., 1246 Heil Quaker Blvd., La Vergne, TN 37086, USA
Lightning Source UK Ltd., Chapter House, Pitfield, Kiln Farm, Milton Keynes, MK11 3LW, GB
BookSurge, 7290 B. Investment Drive, North Charleston, SC 29418, USA
ISBN: 978-3-639-04446-1

Abstract

Resource Allocation (RA) is used to organize the usage of network's physical resources in such a way that guarantees optimal utilization while providing predictive performance to network flows in terms of inter-flow fairness and guaranteed Quality of Service (QoS). One approach to provide RA which is particularly suitable for traffic flows with relaxed QoS requirements (i.e. elastic traffic) is the Utility-based Resource Allocation (URA). URA assigns a utility function to each individual user flow to measure the degree of satisfaction of this user as a result of assigning a specific share of resources. The objective of the URA techniques is then to partition the network resources to take full advantage of them in satisfying the QoS requirements of each user flow while providing fair allocation of resource among users by maximizing the aggregate utility of all flows.

The objective of this thesis is to devise new methods for URA in wired and wireless networks to provide fair resource sharing and predictive flow performance in terms of QoS while guaranteeing best resource utilization. In so doing, we propose a comprehensive set of algorithms that can be used to provide resource optimization both on the link level or on the network level.

On the link level, the thesis proposes a group of algorithms for calculating the optimal classification for a set of traffic flows with diverse QoS requirements to a link with predetermined service levels or predetermined class weights. These algorithms can be used to efficiently study the effect of selecting service levels or class weights according to the distribution of the QoS requirements of the incoming traffic flows. For links with adjustable service levels, we propose two algorithms (**OQP**, and **OQP-OBA**) that calculate the optimal partitioning of traffic flows, the best service levels, and the optimal bandwidth allocation to minimize the quantization overhead as a result of QoS-based partitioning. Our simulation results for both link models show that using 4 or 5 service levels will achieve the trade-off between complexity and service level granularity irrespective of the QoS distributions. These algorithms indeed provide major enhancements in that fairly unexplored area.

On the network level, a key contribution of this thesis is the development of a new decentralized algorithm (**ORAWM**) for resource optimization over multihop wireless networks. The algorithm is used to control the rates of the end-to-end sessions utilizing the bandwidth-efficiency feature of multicast to provide resource optimization in a totally distributed network environment without any synchronization requirements between network node calculations. Through analytical modeling and simulations, we prove the convergence of the asynchronous algorithm under slow network changing conditions such as channel capacity and node mobility. We also devise a detailed network architecture and discuss the protocol implementation for deploying **ORAWM** in an ad hoc network. We also extend our solution to include multicast sessions with heterogeneous receivers

(**ORAHWM**) and discuss the modified network architecture to support multirate multicast trees. The results show that **ORAHWM** not only provides flexibility in allocating resources across multicast sessions, but it also increases the aggregate system utility and improves the overall system throughput by almost 30% compared to homogeneous multicasting (**ORAWM**). We also provide a comprehensive set of simulations that show the effect of deploying these algorithms on the overall resource utilization in an ad hoc network with different environment settings and dynamic network changes (e.g. mobility and route changes).

Contents

List of Tables

List of Figures

List of Abbreviations

ACK	Acknowledgment
AODV	Ad hoc On demand Distance Vector
ATF	Arbitrary Tolerance Factor
BFE	Brute Force Enumeration
DCF	Distributed Coordination Function
Diffserv	Differentiated Services
DP	Dropping Probability
EMH	Extended Multicast Header
FCW	Fixed Class Weights
FSL	Fixed Service Levels
GMPLS	Generalized MPLS
IP	Internet Protocol
LSR	Label Switched Routers
MAC	Medium Access Control
MAODV	Multicast AODV
MCQO	Minimum Class Quantization Overhead

MMP	Multicast aware MAC Protocol
MPLS	Multi-Protocol Label Switching
NS	Network Simulator
OBA	Optimal Bandwidth Allocation
OQC	Optimal QoS-based Classification
OQP	Optimal QoS-based Partitioning
ORAHWM	Optimal Resource Allocation for Heterogeneous Wireless Multicast
ORAWM	Optimal Resource Allocation for Wireless Multicast
PCW	Predetermined Class Weights
PDF	Probability Density Function
PDM	Proportional Differentiation Model
QoS	Quality of Service
RA	Resource Allocation
SL	Service Level
SLA	Service Level Agreement
SP	Service Provider
SRP	Statistical Resource Provisioning
TCP	Transmission Control Protocol
UDP	User Datagram Protocol
URA	Utility-based Resource Allocation
VPN	Virtual Private Network

Dedication

To my beloved parents.

إلى أمى الحبيبة وإلى أبى الحبيب

رَّبِّ ارْحَمْهُمَا كَمَا رَبَّيَانِى صَغِيرًا

Acknowledgements

First and foremost, All praise be to God, for his countless blessings and guidance.

I would like to express my profound gratitude and appreciation to my supervisor, Dr. Hussein Alnuweiri, for his guidance, technical advice, invaluable feedback, understanding, generous support, and friendship.

I would like to thank Dr. Vikram Krishnamurthy and Dr. Rabab Ward for funding my research at some points during the program.

I would like to express my appreciation to my thesis committee for their valuable comments which enhanced the presentation of this thesis.

I am very grateful to Dr. Fayez Gebali from University of Victoria for his valuable discussions regarding the work of QoS partitioning.

I would like to express my sincere thanks to my friend Dr. Watheq El-Kharashi for carefully reviewing my thesis. His valuable comments and suggestions have been very useful in enhancing the presentation of this thesis.

I would like to thank Dr. Yuan Xue from Vanderbilt University for her great help and support throughout the work of resource allocation in wireless multicast and providing the code for the NS implementation which was the starting point of my multicast-based resource allocation for ad hoc networks.

I have been blessed to come in contact with many wonderful people throughout my study period at UBC. This page is definitely not enough to mention all of them, but, the least I can say to them is: thank you for making my stay at UBC such a pleasant experience.

Among the many friends that I had during my study period, the following friends had the most positive effect on my life: Ahmed, Anwar, Ayman, AmrW, MohamedA, Khaled, Tamer, Watheq, Abdo, Junaid, Tariq, Maged, and Awad.

Words cannot describe my feelings toward my parents. I can only pray to God to reward them for their constant support, and persistent encouragement.

I would like to express my love and deep appreciation to the person who stood beside me through all times, my wife Marwa. She was always there for me with comforting words, encouraging thoughts, and a tranquil smile.

I would like to extend my thanks to my brother Mohamed, for being a source of motivation and prayers.

I would like also to thank my in-law's for all their prayers, encouragement, and support.

Last, but not least, I would like to thank my daughters, Sara and Yomna, for all the joy and happiness they have brought to my life.

Chapter 1

Introduction

1.1 Motivation

The rapid growth of customer demands for fast, reliable, and differentiated network services will always go hand in hand with inventing new network infrastructure technologies. On the other hand, service providers (SPs) are striving, not only to fulfill user's demands but also to accommodate more users in order to reduce the service cost without affecting the service quality. This major trade-off has mandated the use of optimization techniques to increase network resource utilization while providing heterogeneous QoS for various types of customer applications.

Problems such as guaranteed QoS or providing differentiated services have proved to be complex to implement on a wide scale despite the great deal of effort that has been dedicated to this subject. Without proper long-term network capacity planning and dynamic mechanisms for allocating and sharing network resources, providing QoS can be an expensive and unsuccessful exercise for network service providers. In recent years, it has become obvious that there is a need for a new paradigm that takes into consideration network wide topology and bandwidth resources to enable SPs to optimize their resource utilization, while maintaining the promised levels of service to customer flows. Proper dynamic resource allocation mechanisms will not only allow SPs to take best advantage of their existing network resources, but will also avoid unreasonable network upgrades as a first choice to respond to ever growing customer demands. Therefore, resource provisioning techniques must be designed to satisfy two primary objectives. The first objective is to satisfy the performance requirements for each customers flow. The second objective is to maximize the overall resource utilization efficiency of the network which in turn impacts the total revenue derived by the SP.

1.2 Network Resource Allocation

Network *resource allocation* refers to the ability of the network to self organize the usage of its physical resources (e.g. bandwidth, buffer space, computation resources, etc.) in such a way that guarantees optimal utilization of these resources while providing predictive performance to individual network flows in terms of inter-flow fairness and guaranteed QoS.

Resource Allocation (RA) can be a powerful tool for exploiting the trade-off between traffic performance from the user side and resource utilization from the network side. From the user side, the job of RA is to provide QoS guarantees such as bandwidth, delay, delay jitter, packet loss etc. From the network side, RA optimizes utilization of network resources to support maximum traffic throughput while providing the required level of QoS.

In packet networks, information is incorporated into the Internet Protocol (IP) packets, and transmitted from one source to one or more destinations in the form of network flows, or often called sessions. One of the most significant and basic network functions is to provide fair share of the network resources to the greatest number of these flows, while efficiently utilizing the resources and guaranteeing an acceptable level of service for each flow. One aspect of achieving this task is to assign individual rates to network flows so that they are able to share the network resources in a fair manner. Without fair allocation, some flows may take over the major part of the network resource while others suffer long network delays or significant data loss. Another important aspect of RA is the repartitioning of network resources to accommodate the flow rates at each network hop and hence guarantee the best resource utilization. Without doing that, some parts of the network may become congested while others may become under utilized, which is highly uneconomic.

1.2.1   Elastic versus non-elastic network traffic

Elastic network traffic [1, 2] usually refers to the network traffic that carries digital objects which can be transmitted using wide range of QoS requirements without affecting the transmission quality. Digital objects such as files, web pages, video clips, or even layered video streams [3, 4] can be transmitted using several possible rates or different QoS levels depending on the limit imposed by the system capacity. In other words, elastic data transmission adapts to available bandwidth via, possibly, feedback control from the network such as in the widely used Transmission Control Protocol (TCP).

Non-elastic traffic (e.g. real time voice or video streams) typically supports hard QoS requirements which must be met by the network in order for this traffic to be admitted. Therefore, it is frequently assumed that elastic traffic has more tolerant QoS requirements compared with non-elastic traffic.

The QoS requirements (e.g. delay, or loss bounds) for both of these types of traffic could be either deterministic or statistical. For example, *deterministic* delay requirements specify the bound on the absolute end-to-end delay for each packet (or over a very short period) while *statistical* delay requirements specify the end-to-end delay bound over a long period possibly with some probability of delay violation.

1.2.2 Resource provisioning for non-elastic traffic

One approach to achieve the compromise between user requirements and network resource utilization which is particularly suitable for non-elastic traffic is the resource provisioning. In this approach, the network resources are partitioned between the individual user flows so as to satisfy the following two objectives:

- The individual QoS requirements for each flow must be satisfied.

- The number of user flows admitted to the network over the long term is maximized.

These two objectives are often conflicting and in some cases it is hard to provide QoS guarantees without over-provisioning network resources for individual flows, which is the main cause of resource under-utilization. Therefore, provisioning techniques frequently opt to provide statistical QoS guarantees [5].

1.2.3   Utility-based Resource Allocation for elastic traffic

One of the most common and efficient methods for modeling the RA problems in packet networks is the concept of *Utility-based* Resource Allocation (URA) for elastic traffic [1, 6, 7]. In this scheme, each network user is assigned a utility function that abstracts the user requirements and measures the degree of satisfaction as a result of assigning a specific share of resources to that user or sometimes it measures the amount of penalty if this user's QoS requirements are not met. In this case the objectives of partitioning network resources are:

- Each network flow acquires a fair share of the network resources that satisfies its QoS requirements to the best degree possible.

- The overall network resource utilization is maximized.

In order to achieve this compromise between fairness and resource utilization, URA techniques are often mapped into an optimization framework that incorporates the user utilities as part of the problem objective. The goal then is to maximize the aggregate utility of all users subject to a set of constraints derived by the limitation of network resources.

Although, it may not be crucial that optimality is exactly attained in real networks, largely due to excessive complexity, the optimization framework offers a means to always steer the network towards a desirable operating point that achieves the trade off between fairness and utilization. Also, mapping the URA problem into an optimization framework has the advantage of leveraging many efficient optimization techniques that can be used to devise and improve solutions that are suitable for distributed computations and can efficiently track network changing conditions in real time.

1.2.4 Per-hop versus end-to-end resource allocation

The resource provisioning and the utility-based resource allocation schemes can address the problem of resource utilization from two levels. First, at the link level, the focus is on providing deterministic/statistical per-hop QoS guarantees for the incoming flows using the proper partitioning of link resources and scheduling flow packets in order to achieve the compromise between fairness and utilization [5, 8, 9]. The notion of end-to-end QoS guarantees is usually not addressed by these techniques. Therefore, resource allocation among per-hop flows may not be suitable for multi-hop flows where packets of the same flow are transmitted across more than one hop . This is due to the unawareness of network-wide bottlenecks and lack of coordination among different hops of the same flow. Second, at the path level, RA techniques focus on partitioning the resources of the entire network in order to provide end-to-end QoS while guaranteeing optimal resource utilization. The proposed solutions in the literature [6, 10, 11] normally use flow control paradigms to perform real-time adjustment on the sending characteristics of the end-to-end flow using feedback from the network.

In this thesis, we focus on the analysis and design of utility-based resource allocation techniques that can achieve optimal resource allocation both at the link (per-hop) level and at the path (end-to-end) level. We propose a comprehensive set of resource allocation algorithms that provide efficient partitioning of system resources, QoS-based classification of flows to these resources, and system-wide fairness for elastic network traffic. Details of the problems, and our proposed solutions will be presented in later chapters. At the link level, we focus on partitioning the link resources for a group of flows to obtain the optimal resource utilization while providing per-hop service granularity. This objective is reasonable for wired networks, especially for an accumulative QoS metric (e.g. dropping probability) since providing service granularity for one-hop flows will help meet the end-to-end QoS requirements. On the other hand, for multihop wireless networks, achieving certain criteria like fairness among single-hop flows may not be optimal for multi-hop flows, due to the challenging features including unpredictable channel behavior, location based contention, mobility, and route changes. Therefore, for multi-hop wireless networks, we focus on allocating rates (QoS metric) across the entire network to achieve optimal utilization while providing fairness to the end-to-end flows.

1.3 The Thesis

1.3.1 Objectives

The main objective of this thesis is to develop new methods for resource allocation in wired and wireless networks that provide fair resource sharing and predictive flow performance while guaranteeing best resource utilization.

1.4 Contributions

The contributions of this thesis can be highlighted as follows:

- Proposal of a new set of algorithms for QoS-based classification for multi-class link models. Our algorithms, based on dynamic programming, will be explained in Chapter 3. This work is published or to be published in [12] [13] [14] .

- Proposal of a new set of algorithms for QoS-based partitioning of link resources to provide resource allocation. Our algorithms will be explained in Chapter 4. This work is published or to be published in [17].

- Development of a new *decentralized* algorithm for end-to-end resource allocation in homogeneous wireless multicast-aware ad hoc networks. Our algorithm (**ORAWM**) and a detailed optimization model based on gradient projection, and distributed computation methods will be explained in Chapter 5. This work is published or to be published in [18] [19] [20].

- Design of a cross layer framework that realizes the problem of end-to-end resource allocation in wireless multicast. This framework utilizes a measurement-based technique for wireless channel capacity estimation and a light-weight network HELLO protocol for constructing contention domains. Our framework will also be explained in Chapter 5. This work is published or to be published in [21] [20].

- Development of a new *decentralized* algorithm for end-to-end resource allocation in heterogeneous wireless multicast-aware ad hoc networks. Our decentralized algorithm (**ORAHWM**) and a detailed optimization framework will be explained in Chapter 6. This work is published or to be published in [22] .

1.5 Thesis Methodology and Road Map

For each of the contributions listed above, we devise a framework that realizes the resource allocation problem as a non-linear optimization model, and we provide the analysis and the optimal solution using one of the most efficient and suitable techniques for optimizing the objective in this case. We also devise heuristic techniques that can be used to approximate the solution and provide analytical and simulation studies to compare these heuristic solutions with the optimal one.

The rest of the thesis is organized as follows: Chapter 2 covers the literature review and provides an overview about the resource allocation techniques in wired and wireless networks.

Chapter 3 proposes a set of algorithms for QoS-based classification for link models with limited number of classes of service.

Chapter 4 Proposes the algorithms for QoS-based partitioning and bandwidth allocation for link models with adjustable service levels.

Chapter 5 provides a comprehensive solution for optimal resource allocation for homogeneous wireless multicast (**ORAWM**) which is based on distributed computations, and iterative techniques.

Chapter 6 provides the solution for optimal resource allocation for heterogeneous wireless multicast (**ORAHWM**).

Chapter 7 concludes this thesis and outlines future research directions.

1.6   Summary

This chapter has covered an introduction to resource allocation in packet networks, and highlighted two solution scenarios for achieving the trade-off between providing QoS guarantees and attaining the best resource utilization, namely: resource provisioning for non-elastic traffic, and utility-based resource allocation for elastic traffic. It has also highlighted the main differences between elastic, versus non-elastic traffic, and per-hop versus end-to-end QoS guarantees in developing resource allocation techniques in packet networks. The last sections of the chapter have covered objectives, contributions, methodology and road map for this thesis.

Chapter 2

Resource Allocation in Wired and Wireless Networks

2.1 Introduction

In today's networks one of the main goals of a service provider is to satisfy their customer demands while using network resources efficiently. Some customer applications with hard QoS requirements require sophisticated QoS provisioning techniques to partition network resources in order to provide hard QoS guarantees for such applications. In some cases, the provisioning technique may have to over-estimate the amount of resources allocated to each customer flow to provide guaranteed QoS, which might lead to under-utilization of network resources. Other customer applications with relaxed QoS requirements require resource allocation techniques that guarantee different notions of fairness (e.g. proportional fairness, maxmin fairness [24, 25, 26]) among customer flows and maximize network resource utilization.

2.2 Per-hop resource allocation in wired networks

Several techniques have been proposed in the area of providing per-hop resource allocation for wired networks using scheduling and bandwidth sharing mechanisms [23, 27, 28,
29, 30]. Some of these techniques have focused on providing per-flow QoS guarantees
while using the link capacity as a continuous set of rates that can be allocated freely to
different flows [31]. These techniques adopt different policies for allocating link capacity into different flows so that the overall link utilization is maximized. Several other
techniques provide differentiated services for aggregated flows on a single link that has a
discrete set of predetermined QoS service levels [32, 33, 34]. The idea of the DiffServ architecture [35] is to provide a scalable solution to the problem of service differentiation by
maintaining the state of only the aggregates of flows. However, the DiffServ architecture
and techniques do not provide answers as to what are the best set of QoS service levels
supported on each link that can achieve maximum resource utilization. Determining the
number and the nature of service levels that each link can support are important issues
that can lead to enhancing the link utilization. Increasing the number of QoS Service
levels on each link will increase the overhead of flow management and classification while
maintaining small number of service levels may lead to waste of link resources. Also
certain applications may inherently restrict the aggregation of traffic flows and therefore
require special per-flow guarantees using limited set of link service levels [8, 36, 37]. For
example, multicast flows and some secured (e.g. VPN flows) trunks may impose some
aggregation challenges at different parts of the network. For this type of flows, selecting

the discrete set of service levels supported by the link is a crucial issue for enhancing the link's resource utilization. This is because assigning a service level that provides better QoS than the one required by the flow will lead to waste of resources while, assigning a service level with worse QoS may not be accepted by the flow or the service agreement with the customer.

Although per-hop resource allocation techniques have wide acceptance and they are wide spread across many of today's networks, they lack the concept of optimizing the entire network resources for multihop flows. This is because fairness of allocating resources among flows on one hop may not achieve the overall fairness across multihop flows.

2.3   End-to-end resource allocation in multi-hop networks

End-to-end resource allocation techniques are designed to achieve global optimal resource utilization for the entire network while providing system-wide fairness among the incoming flows. Most of these techniques use flow control, sometimes combined with topology-aware congestion mechanisms, to perform real-time adjustment on the sending characteristics (e.g. sending rate) of the end-to-end flow. The purpose of flow control is to regulate the admission of traffic into the network in order to minimize network overload, without excessively restraining sessions when the network is not overloaded. Flow control techniques are usually divided into two categories, namely: window-based, and rate-based flow control.

Window-based flow control techniques [38, 39] use a limit on the maximum number of unacknowledged packets submitted to the network at any given time (the window size). They rely mainly on the time taken by the per-packet acknowledgment to be fed back to the source in estimating the congestion state of the network, and hence they use that time in controlling the sending rate of the flow source. If the network is highly congested, acknowledgments will take longer time to reach the source and that will trigger the source to slow down by reducing its sending rate. While these techniques are intuitively simple to implement, they usually lack the coordination between the competing flows and therefore they provide poor inter-flow fairness. Besides, without using a network resource assignment scheme, window-based techniques do not provide deterministic guarantees neither on the network resource utilization nor on the end-to-end QoS of each flow.

Rate-based flow control techniques on the other hand, use a combination of rate adjustment and resource assignment functions to control the characteristics of the flows. The resource assignment function of the flow control technique divides the available resources (e.g. link or channel capacity) amongst flows to ensure that the QoS requested by each flow is fulfilled and resource utilization is maximized. Of course in that case, network topology plays an important role in maximizing the resource utilization across the entire network. Also to maintain scalability and minimize the coordination between network nodes, the majority of the topology-aware resource assignment mechanisms are based on distributed solutions which reduce the computations performed by each node and enhance the communication overhead. The rate adjustment function controls the sending rate of each end-to-end flow based on feedback from the network to guarantee inter-flow fairness. Because of the complexity of achieving global optimal resource utilization and topology awareness, rate-based techniques are normally designed for a special type of network (e.g. optical, wireless, etc.) and for a special type of end-to-end flows (e.g. unicast, multicast flows etc.).

2.3.1 End-to-end resource allocation in wired networks

In this class of end-to-end resource allocation techniques, the link between two network nodes represents the main resource entity for which flows at the network router contend for resources. Scheduling at each individual link regulates the provisioning of link bandwidth to each flow (i.e. time domain packet multiplexing).

The problem of optimal and fair resource allocation has been extensively studied in the context of wired networks for unicast flows. Some techniques focused on providing end-to-end QoS guarantees while maximizing the number of admitted sessions by partitioning the end-to-end QoS requirements of each flow into local QoS on each network link based on the link's loading conditions [40, 41]. Other techniques have focused on providing optimal resource utilization while guaranteeing inter-flow fairness. These techniques adopt a pricing scheme to design a distributed solution for rate allocation for unicast flows [1, 10, 42]. They use a local congestion parameter (the price) to convey the information about the relationship between the capacity on each link and the load demanded by the flows and hence provide a congestion indication for each part of the network.

While resource allocation for unicast flows focuses on assigning resources to each source-destination pair, multicast flows present more challenge because of the correlation between the source-destination pairs on the same multicast tree. Such correlation requires special treatment especially for the case of multirate multicast [43]. Resource allocation for multicast wired networks has been studied in [44], [45] and [7]. They define the multicast tree between one source and multiple receivers as group of terminal (non-junction) nodes and branch (junction) nodes. Each downstream on a branch node contribute to the load of the link that the downstream traverses. They also associate a utility function to each receiver on a multicast session to reflect the demand for bandwidth for this receiver on each multicast session. The objective then is to find the allocation of resources across the multicast trees such that the aggregate utility of all receivers on all session is maximized, taking into consideration the correlation between the source-destination pairs of each multicast tree.

2.3.2   End-to-end resource allocation in multihop wireless nctworks

In this class of end-to-end resource allocation techniques, wireless channel represents the main resource entity for which flows at the network nodes contend. All network nodes within the interference range of each other constitute a cluster of nodes for which only one node can send packet at any given time. Therefore, scheduling techniques for such wireless network regulate the access of the flows to the wireless channel using distributed coordination between the interfering nodes both in time and spatial domains. Besides, unpredictable channel behavior caused by impairments, such as channel noise and interference, introduces additional challenge that makes provisioning of channel bandwidth such a difficult job especially for providing QoS guarantees.

Resource allocation, using MAC-layer fair scheduling for single-hop flows, has been studied in wireless ad hoc networks [46, 47, 48]. Such techniques, however, do not provide end-to-end resource optimization and therefore fairness among the multihop flows cannot be achieved only by local MAC layer scheduling. End-to-end resource allocation techniques (e.g. [6, 49, 50, 51]) complement such MAC layer scheduling techniques by providing a pricing mechanism that can be used to coordinate the global resource allocation for the case of multihop unicast flows. In this case, a utility is associated to each end-to-end unicast flow, and the objective is to maximize the aggregate utility subject to channel capacity limitations. Global optimality has also be addressed in the context of maximizing the net benefit of each node as a result of relaying traffic for other nodes [52, 53]. There too, a utility is assigned to each user, and the objective is to maximize the aggregate profit of all nodes within the network subject to the limited resources of each node. Another context where the problem of global resource allocation has been addressed is the distributed power-based resource allocation [54, 55, 56, 57, 58]. In this case, a similar pricing mechanism is used to optimize the signal-to-interference (SIR) ratio using a distributed power control technique. None of these techniques, however, investigated the case of end-to-end multicast flows.

The one hop broadcast characteristic of the MAC layer in wireless ad hoc networks has prompted the use of multicast communication scheme as one of the natural strategies that can multiply the overall network throughput with very limited overhead. This is because multicast packets are forwarded once to reach all the multicast members in the neighborhood using a single transmission, and this effect increases even more in multi-hop ad hoc networks. However, historically, multicast flows introduce new challenges for deployment due to its complexity and lack of standards and techniques that can effectively allocate network resources. The challenge in multicast resource allocation is that, not only the resource allocation has to regulate access to channel resources for each source-destination pair, but also it has to correlate different source-destination pairs on the same multicast group. This is crucial for utilizing the bandwidth efficiency of each multicast session (i.e. "intra-flow bandwidth efficiency") and achieving system-wide fairness across all multicast sessions (i.e. inter-flow fairness).

2.4   Summary

This chapter has covered a literature review for resource allocation in wired and wireless networks. First, a survey on single-hop based resource allocation techniques which are designed for wired networks has been presented. Then, the latest techniques for end-to-end resource allocation for multihop networks has then been discussed including the motivation for our multicast resource allocation in multihop wireless ad hoc networks.

Chapter 3

Optimal QoS-based Classification for Multi-class Link Models

In this chapter, we present a comprehensive set of algorithms for QoS-based classification for a multi-class link model with a small predetermined set of service levels. We present two main algorithms called **OQC-FSL** and **OQC-PCW** for calculating the optimal QoS-based classification using dropping probability as the QoS metric. We also provide a comparison between these algorithms and two heuristic techniques in calculating the quantization overhead resulting from the classification process.

3.1 Introduction

The proliferation of differentiated services has been one of rising challenges to Internet SPs aiming to deploy cost-effective, large scale networks, supporting diverse applications and services ranging from web browsing to video-on-demand and TV broadcast. On the other hand, differentiated services promote the idea of supporting limited service levels

---
[0]A version of this chapter has been published [12, 13, 14]

to provide QoS while maintaining the network scalability. This idea has wide range of acceptance by today's applications. However, some emerging applications and services such as TV-over-DSL, virtual private networks (VPNs), overlay networks, and multicast-based QoS networks, may inherently restrict the aggregation of traffic streams in network nodes. Nonetheless, they still require different levels of guarantees at each network node in order to guarantee the end-to-end QoS. Therefore, the problem of satisfying per-flow QoS guarantees in a network environment that supports only a fixed set of service levels and traffic classes has been the subject of extensive research [8, 36, 37].

To provide the required end-to-end QoS, many resource reservation mechanisms have been discussed in the literature and can be used to partition the end-to-end QoS into local QoS requirements on each network element (link) [40, 59, 60]. However, for a network framework that supports a limited set of service levels on each network link, the local QoS required for each connection (or flow) will be quantized based on these service levels. Assigning a service level which provides better QoS than the one required by the connection will lead to waste of network resources while, assigning a service level with worse QoS may not be accepted by the application or the service agreement with the network user. Therefore, an important network optimization problem is how to assign the network connections to the best service level that minimizes the total quantization overhead. Such an overhead is the quantization penalty that might have to be incurred by either the user or the service provider. Hence, addressing this problem will have a large significance especially for high speed links (e.g. optical links) that may potentially carry large number of connections.

In this chapter, we investigate the problem of optimal classification of connection requests into a set of service levels at a network link in order to minimize the penalty caused by QoS quantization. We consider a model where each connection request has a local QoS (e.g. delay, dropping probability, jitter, etc.) in addition to the bandwidth requirements. The link has a set of service levels that are either fixed or defined by a set of predetermined class weights. Each connection request is assigned to one of the link's service levels that best accommodates the connection QoS requirements, such that the total QoS quantization penalty is minimized.

In Section 3.2, we define a utility function which will be used to measure the quantization penalty at each service level based on the set of connection requests assigned to that level. We also define a set of conditions on this utility function that will be used by the call admission process to accept or reject new connection requests. We present two categories of efficient (polynomial time) algorithms which can be used for fixed service levels and for predetermined class weights given some realistic assumptions.

The problem of obtaining optimal quantized service levels for a set of connection requests has been investigated before in [61], [4], and [62]. Rouskas et al. [61] defined a similar framework of assigning MPLS tunnels with specified data rates to a set of quantized service levels such that the performance penalty of wasted bandwidth is minimal. Both the deterministic and stochastic connection requests have been discussed. However, the case of having two joint requirements (e.g. Bandwidth, and dropping probability) was not discussed. The authors in [4], and [62] considered the problem in a different context where the set of requests are group of receivers on a multicast tree. The target then was to find the small optimal set of offered rates (by the source) that maximizes the correlation between the total required rates requested by the receivers and the offered service levels. The present study offers solutions to the generalized problem where connections have both rate and QoS requirements and presents polynomial time algorithms for fixed service levels or predetermined class weights link model for soft and hard QoS requirements.

The rest of the chapter is organized as follows. Section 3.2 formulates the model, the problems of QoS classification and relates our framework to QoS network architectures. The optimal solution for fixed service levels is discussed in Section 3.3. The optimal solution for admission control is investigated in Section 3.4. Sections 3.5 and 3.6 discuss the case of predetermined class weights and describe how the optimal solutions will be modified in this case. Section 3.8 depicts the results for applying our solutions to a set of QoS requirements with different distributions. Finally, concluding remarks and future work are described in Section 3.9.

3.2 Model and problem formulation

3.2.1 Model

We assume a network element, or a link of capacity C that offers a set of service levels $S_L = \{x_1, x_2, ..., x_L\}$ for L number of classes. Typically, the values of the service levels are limited by the link capacity and dependent on the inherent link model and the characteristics of the input traffic. The *l-th* level corresponds to a connection with QoS level x_l given that the bandwidth demand for this connection is $\leq C$. Similarly, we assume a set of N connection requests $\Re = \{r_1, r_2, ..., r_N\}$, each defined by a bandwidth demand parameter d_i plus one local QoS requirement Q_i (e.g. average delay, dropping probability, jitter, etc.). Therefore, each connection request can be represented by a pair, viz. $r_i \equiv \{d_i, Q_i\}$ as shown in Figure 3.1.

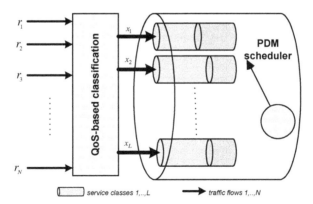

Figure 3.1: Link Model.

It is important to note the difference between parameters x_l and Q_i. Essentially, x_l is the actual QoS achieved at service class l, while Q_i is the original (or absolute) QoS requirement of connection $r_i \equiv \{d_i, Q_i\}$. The classification process assigns connections from the set $\Re = \{r_1, r_2, ..., r_N\}$ to service levels from the set $S_L = \{x_1, x_2, ..., x_L\}$ using mapping functions that will be explained in the next section. As a result of the assignment process, a connection request may be assigned to a service class that provides the exact required QoS, inferior QoS, or better QoS to the connection. Unless otherwise stated, for the remaining of this chapter, if $x_l < Q_i$, this means connection i is assigned a *better* QoS (note that this will not be true if Q_i's represent bandwidth or throughput for example). The difference between the absolute QoS and the achieved QoS, is an important measure that will be used to evaluate different QoS classification schemes. In the following, we develop the definition of a *utility function*, $U(Q_i, x_l)$, which measures the difference between the achieved and required (or absolute) QoS. It is important to note that providing x_l lower than Q_i (i.e. providing better QoS) will result in a waste of allocated resources whereas increasing x_l over Q_i may have a disadvantage for the application generating the traffic stream. Therefore, we will consider the general case where $U(Q_i, x_l)$ measures the penalty in both scenarios (different from the one defined in [61]). The following are some common examples from the literature.

The absolute difference

$$U(Q_i, x_l) = |Q_i - x_l| \tag{3.1}$$

The least square

$$U(Q_i, x_l) = (Q_i - x_l)^2 \tag{3.2}$$

The logarithmic difference

$$U(Q_i, x_l) = \log(1 + |Q_i - x_l|) \tag{3.3}$$

We consider a subset of the possible classification policies where the connection requests are sorted based on the local QoS requirements in non-decreasing order, (i.e. $Q_1 \leq Q_2 \leq Q_3 ...... \leq Q_N$). This is called ordered classification.

Definitions

Ordered Classification: A group set $\pi_L = \{G_1, .., G_L\}$ is an *ordered classification* if $\forall r_i \in G_a$ and $\forall r_j \in G_b$, $Q_i \leq Q_j$, when $a < b$.

Utility Property: Intuitively, the utility function is non-increasing in the interval $Q_i \in [0, x_l]$ and it is non-decreasing in the interval $Q_i \in [x_l, \infty[$. In other words, the utility property can be defined using the following inequalities:

1. $\forall x, i = 1, .., N \quad j = 1, .., L$

   if $Q_i < Q_j \leq x$ then $U(Q_i, x) \geq U(Q_j, x)$

   if $x \leq Q_i < Q_j$ then $U(Q_i, x) \leq U(Q_j, x)$

2. $\forall Q, i = 1, .., N \quad j = 1, .., L$

   if $x_i < x_j \leq Q$ then $U(Q, x_i) \geq U(Q, x_j)$

   if $Q \leq x_i < x_j$ then $U(Q, x_i) \leq U(Q, x_j)$

The importance of the last two definitions lies in the fact that if we choose a utility function that fulfills the utility property we only need to consider the ordered classifications when we search for an optimal solution. This result is stated formally in the following fundamental lemma.

Lemma 3.1. *For any utility function* $U(Q_i, x_l)$ *that satisfies the utility property, even with predetermined service levels, there always exists a classification* π_L^* *that is optimal and ordered.*

Proof. Given in Appendix A   □

In section 3.5, we will impose one more restriction on the utility function which will lead to an elegant solution for the case of predetermined class weights.

Lemma 3.1 is important because it limits the number of possible solutions for optimality, making the problem much more tractable. The following lemma highlights this fact and gives a deterministic value to the number of possible solutions for the ordered classification problem.

Lemma 3.2. *Brute Force Enumeration (BFE) of input streams using the ordered classification policy will lead to a number of possible solutions in the order of* $O(N^{L-1})$ *where* N *is the number of traffic streams and* L *is the number of service levels.*

Proof. Given in Appendix A   □

Although, the lemma indicates that BFE for the input traffic streams will lead to an algorithm with exponential running time in L, we can argue that typical values for L are normally small and the BFE method will in fact lead to a polynomial-time solution for a fixed value of L. In the following sections, we will show that the recommended value for L is, remarkably, small and in the range of 3 to 5 classes. We will also present our solutions which achieve polynomial running time regardless of the value of L.

3.2.2 Optimization Problem

In general, the optimization problem of concern here is to classify N connection requests $\Re = \{r_1, r_2, ..., r_N\}$ to one of the L service levels $S_L = \{x_1, x_2, ..., x_L\}$ while minimizing the quantization overhead. Figure 3.2 illustrates the classification process for $L=4$ service levels. Each point in the figure corresponds to a connection request with a QoS and bandwidth requirement. Vertical lines represent the separation between the service levels. Formally, we need to find the set of L classified groups (classes) $\pi_L^* = \{G_1^*, G_2^*, ..., G_L^*\}$ such that $\psi(\pi_L^*) \leq \psi(\pi_L) \quad \forall \pi_L$ where $\psi(\pi_L)$ measures the service quantization overhead and is defined as follows:

$$\psi(\pi_L) = \sum_{l=1}^{L} \sum_{\forall r_i \in G_l} U(Q_i, x_l) \tag{3.4}$$

Each connection request has to be assigned to exactly one service level. Also, because each service level must have some resources assigned to it (e.g. buffer, bandwidth, channel, etc.), we will restrict the groups G_l $l=1,...,L$ to be non-empty. In other words, the size of each group $|G_l| \geq 1$, $l=1,...,L$.

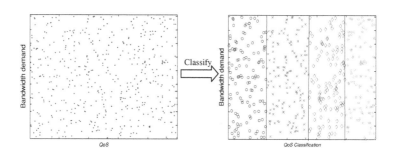

Figure 3.2: QoS-based classification.

3.2.3 Proportional Differentiation Model (PDM)

We consider a link that satisfies the proportional differentiation model (PDM) with a fixed set of weights for a finite number of traffic classes. Such a model has been adopted by several proposed mechanisms in different types of networks like [34], [33], and [32]. The PDM dictates the following relationship for all pairs of service classes.

$$\frac{x_i(t, t + \tau)}{x_j(t, t + \tau)} = \frac{\omega_i}{\omega_j} \tag{3.5}$$

where x_i, and x_j are the service levels (e.g. average delay, dropping probability, etc.) achieved for two classes i and j over the time period τ, and ω_i and ω_j are the weights assigned to these classes. We consider connection requests with life time long enough for Equation (3.5) to be valid.

Example Link Model

We consider for example the model used in [34]. The statistical model for Dropping Probability (DP) is explained in [63]. The DP for Poisson traffic arrival is defined as

$$DP = \frac{\rho^{m+B}}{m!m^B[1 + \sum\limits_{i=1}^{m} \rho^i\frac{1}{i!}] + \sum\limits_{i=m+1}^{m+B} \rho^i m^{m+B-i}} \tag{3.6}$$

where ρ is the link's total load (i.e. $\rho = \sum\limits_{i=1}^{N} d_i/C$), m is the number of data channels, and B is the number of storage locations (in packets). Furthermore, if we divide this link into L classes where all classes are active, and traffic in all pairs of classes are statistically independent, the relationship between the total DP and x_l values is as follows:

$$DP = 1 - \prod_{l=1}^{L}(1 - x_l) \quad \Rightarrow \quad -\log(1 - DP) = -\sum_{l=1}^{L}\log(1 - x_l)$$

Considering that x_l's are $<<1$ (i.e. $-\log(1 - x_l) \approx x_l$), then

$$-\log(1 - DP) \approx \sum_{l=1}^{L} x_l \tag{3.7}$$

From Equations (3.5), (3.6), and (3.7), we can obtain the link's offered service levels for Poisson traffic. The same approach can be used with other dropping probability models for Poisson and voice traffic models such as those suggested in [64] and [65].

Example

Assume we have 5 connection requests having a total bandwidth demand of 110Mbps, with the following set of DPs $Q= \{0.00033027, 0.0036963, 0.024236, 0.043694, 0.074222\}$. Assume that we need to assign the 5 requests to 3 traffic classes with the following weights

$W = \{10^{-3}, 10^{-2}, 10^{-1}\}$, given a link capacity of 100Mbps. In general, we have 6 possible ordered classifications for the 5 connections.

We can obtain the service levels using the model explained before (ρ=1.1, m=1, B =200), $S_L = \{0.000859, 0.008586, 0.085865\}$. We can then obtain the optimal classification using these service levels as will be explained in Section 3.3.

Optimal classification $\pi_L^* = \{[0.00033027, 0.0036963], [0.024236, 0.043694], [0.074222]\}$ and the total quantization overhead in this case is 0.065766 (using $U(Q_i, x_l) = |Q_i - x_l|$).

- It allows us to develop methods for optimally classifying general traffic, which are scalable because they have no shaping restrictions.

- It will have the highly desirable feature of decoupling bandwidth from QoS guarantees. This is not the case with most packet schedulers.

3.3   Optimal solution for fixed set of service levels and soft QoS requirements

For a set of service levels $S_L = \{x_1, x_2, ..., x_L\}$ which may be based on the link's total load and a set of QoS requests $\varphi_N = \{Q_1, Q_2, ..., Q_N\}$, we form the system matrix $A_{N \times L}$ such that the elements of A are $a_{ij} = U(Q_i, x_j)$ i=1,...,N, and j=1,...,L.

To minimize $\psi(\pi_L)$, we map the problem to the following dynamic program.

$$\psi(i, l) = \min_{l-1 < k < i-1} [\psi(k, l-1) + \sum_{j=k+1}^{i} a_{jl}]$$

where $\psi(i,l)$ measures the minimum quantization overhead for the l-*th* service level and up to the i-*th* traffic stream recursively based on the overhead of previous service levels. Based on this equation, we propose an algorithm for finding the optimal QoS-based classification for a fixed service levels **OQC-FSL** shown in Figure 3.3 for the classification problem. Note that the summation of a_{jl} for $j = k+1,\ldots,i$ can be pre-computed for all $i=1,\ldots,N$ and $j=1,\ldots,L$ leading to a complexity of $O(N^2L)$. The matrix CG is used by the **Dynamic_Program** sub-algorithm to store the indices of the current best classified group for each service level.

Lines 1 and 2 of the main algorithm **OQC-FSL** have complexity of $O(N^2)$ and $O(NL)$ respectively in the case that the QoS values are not sorted. The pre-computation complexity for the summation of a_{jl} for $j = k+1,\ldots,i$ for $i,j=1,\ldots,N$ and $k=1,\ldots,L$ in line 6 of the **Dynamic_Program** sub-algorithm is $O(N^2L)$ which is similar to the complexity of the **Dynamic_Program**. Therefore, the complexity of algorithm **OQC-FSL** is $O(N^2L)$.

3.4   Optimal solution in the context of admission control (hard QoS Requirements)

Until now, we have assumed that incoming traffic streams can tolerate service levels with a QoS value lower than the QoS it required initially. Although many practical applications can tolerate certain amount of QoS level drops, the level of tolerance can

$$(\psi(\pi_L^*), \pi_L^*) = \textbf{OQC-FSL} \; (\{Q_1, ..., Q_N\}, \{x_1, ..., x_L\})$$

1. $Q_i \quad \forall i$ are sorted in a non-decreasing order

2. Form matrix $A_{N \times L} \equiv \{a_{ij} \; i = 1, ..., N \quad j = 1, ..., L\}$

3. $(\psi(\pi_L^*), \pi_L^*) = \textbf{Dynamic\_Program} \; (A)$

$$(\psi(\pi_L^*), \pi_L^*) = \textbf{Dynamic\_Program} \; (A)$$

1. **For** $l = 1$ to L

2.   **For** $i = 1$ to N

3. $\psi(i, l) = \infty, \quad CG(i, l) = 0$

4.   **For** $k = l - 1$ to $i - 1$

5.   **If** $\psi(k, l - 1) + \sum\limits_{j=k+1}^{i} a_{jl} < \psi(i, l)$ **Then**

6.   **Set** $\psi(i, l) = \psi(k, l - 1) + \sum\limits_{j=k+1}^{i} a_{jl}, \;$ and $CG(i, l) = j$

7.   **End If**

8.   **End For**

9.   **End For**

10. **End For**

11. $\psi(\pi_L^*) = \psi(N, L)$

12. **For** $m = L$ **Down To** 1

13.   **Set** $G_m^* \in \pi_L^* = \{r_i, \; i = (CG(j, m) + 1), ..., j\}$

14.   **Set** $j = CG(j, m)$

15. **End For**

Figure 3.3: Optimal QoS-based classification with fixed service levels (OQC-FSL).

vary widely among different applications. Therefore, an *admission control* mechanism
is needed to impose some restrictions on the minimum level of service achieved by all

traffic streams (the new added steams plus the existing ones). Basically, to admit new streams, the link must be able to provide all traffic streams with a level of service which is within a certain tolerance limit from their required QoS. Otherwise, the admission control mechanism might reject the addition of new traffic streams. In order to capture this behavior, we introduce a new variable called the *tolerance factor, δ*. This factor represents the maximum drop in QoS for the traffic stream that can be tolerated by the application generating this traffic. For example, in streaming applications the tolerance factor can be improved by adjusting the size of the play-back buffer at the receiver. The tolerance factor can be part of the service level agreement (SLA). Note that $\delta=0$ means that the traffic stream cannot tolerate any service level drop. We will assume the most general case in which each traffic stream has a different tolerance factor (i.e. $r_i \equiv \{d_i, Q_i, \delta_i\}$). To accommodate this fact, we will add the following condition:

$$(Q_i - x_j + \delta_i) \geq 0 \quad i = 1, ..., N \; j = 1, ..., L$$

Or if δ_i is defined as a *percentage* of QoS Q_i, the condition becomes

$$((1 + \delta_i) \, Q_i - x_j) \geq 0 \quad i = 1, ..., N \; j = 1, ..., L \tag{3.8}$$

To incorporate this condition in the **OQC-FSL** algorithm we will exclude the assignments where condition (3.8) is not satisfied by setting $a_{ij} = \infty$ as shown in Figure 3.4. This will force the **Dynamic_Program** to assign traffic streams only to the classes that can provide service levels within the tolerance factor.

$$(\psi(\pi_L^*), \pi_L^*) = \textbf{OQC-FSL-ATF} \ (\{\Re\}, \{x_1, ..., x_L\})$$

1. $Q_i \quad \forall i$ are sorted in a non-decreasing order

2. Form matrix $A_{N \times L} \equiv \{a_{ij} \ i = 1, ..., N \quad j = 1,...,L\}$

3. **If** $((1 + \delta_i) \ Q_i - x_j) \ \pi_{ij} < 0$ **Then Set** $a_{ij} = \infty$

4. $(\psi(\pi_L^*), \pi_L^*) = \textbf{Dynamic\_Program} \ (A)$

Figure 3.4: OQC-FSL with arbitrary tolerance factors (OQC-FSL-ATF).

3.5 Service differentiation based on class weights

We start by assuming that the maximum link capacity restriction is not enforced or, alternatively, assume that there is a light traffic load on the link. Obviously, if the link is lightly loaded, the QoS achieved by the traffic streams may potentially exceed the QoS requirements. In this case, we have some flexibility in assigning the traffic streams to the different service classes. One scheme would be to assign most of the traffic streams to the lowest service level(s) as long as the QoS requirements by the streams are satisfied. However, even in this case, it may be desirable to offer some service differentiation between traffic streams while saving the link resources, for example, to increase bandwidth available for best effort traffic. To achieve that, we follow the same methodology explained in Section 3.3. This time, we set $a_{ij} = U(Q_i, \gamma \omega_j)$ such that ω_j is the weight assigned to service class j, $j=1,...,L$, and γ is the service differentiation factor selected based on the characteristics of incoming QoS requests.

By choosing a specific value for γ, the problem reduces to fixed set of service levels which was explained in section 3.3. Ideally, we need to select the value of γ such that

$\psi(\pi_L^*)$ has the lowest value possible. However, this means that we need to try all possible values of γ and get the optimal classification for each case, then select γ such that the value of $\psi(\pi_L^*)$ is minimized. Such naïve approach has very high complexity. In the following, we propose two heuristic estimates for the value of γ. Then, we drive the optimal value of γ that guarantees a minimum value for $\psi(\pi_L^*)$ $\forall \gamma$. First, we present the following lemma.

Lemma 3.3. *If $U(Q_i, \gamma\omega_j)$ fulfills the utility property, and $U(Q_i, \gamma\omega_j) = C_{ij}*U(Q_i/\omega_j, \gamma)$, where C_{ij} is any constant, then in order to minimize the value of $\psi(\pi_L^*)$, γ must be selected in the range $[\min(Q_i/\omega_j)$, $\max(Q_i/\omega_j)]$ $i = 1, ..., N$ $j = 1, ..., L$.*

Proof. Given in Appendix A   □

This lemma limits the range of possible values for γ between the minimum and maximum possible weighted QoS values.

Using *median* as a heuristic estimate of γ

One natural choice for the value of γ is the center of the sample data $d_{ij} = Q_i/\omega_j$, or the statistical **median**(d_{ij}) $i = 1, ..., N$ and $j = 1, ..., L$. The intuition behind this selection is to try to lower the values of $U(Q_i, \gamma\omega_j)$ by selecting γ in the center of d_{ij} in order to minimize the value of $\psi(\pi_L^*)$. The complexity of calculating the statistical median is known to be $O(NL)$[66].

Using *minimum summation* as a heuristic for calculating γ

Another heuristic estimate is to select the value of γ such that the total $\sum_{i=1}^{N} \sum_{j=1}^{L} a_{ij}$ is minimized. The intuition is that if we minimize this summation, then we may potentially lower the value of the sum on any subset of i, and j. Although in this case, we have an infinite number of values for γ to try, the following theorem limits the possible choices of γ when some reasonable conditions on the utility function are satisfied.

Theorem 3.1. *If $U(Q_i, \gamma\omega_j)$ is a piecewise concave function, and $U(Q_i, \gamma\omega_j) = C_{ij} * U(Q_i/\omega_j, \gamma)$ where C_{ij} is any constant, then γ must take one of the values of $d_{ij} = Q_i/\omega_j$ $i = 1, ..., N$ and $j = 1, ..., L$ in order to minimize the summation $\sum_{i \in I} \sum_{j \in J} a_{ij}$ for any subset of i and j.*

Proof. Given in Appendix A   □

The above theorem implies that to get the value of γ which minimizes the summation $\Psi = \sum_{i=1}^{N} \sum_{j=1}^{L} a_{ij}$, we need only to consider $N*L$ possible values of $\gamma \in \{d_{ij} = Q_i/\omega_j, \; i = 1, ..., N \; j = 1, ..., L\}$, given that $U(Q_i, \gamma\omega_j)$ is a piecewise concave function and the condition $U(Q_i, \gamma\omega_j) = C_{ij} * U(Q_i/\omega_j, \gamma)$ is satisfied. Because it is a piecewise concave function, $U(Q_i, \gamma\omega_j)$ is non-increasing concave in the interval $Q_i \in [0, \; \gamma\omega_j]$ and non-decreasing concave in the interval $Q_i \in [\gamma\omega_j, \infty[$. Some examples for piecewise concave functions are defined by Equations (3.1) and (3.3).

To illustrate the effect of Theorem 3.1, Figure 3.5 shows the summation Ψ for all possible values of γ using $U(Q_i, \gamma\omega_j)$ as defined by Equation (3.3) for three incoming traffic streams and one class of service. It is clear from the figure, that the possible minimum values of Ψ correspond to $\gamma \in \{Q_1/\omega_1, \; Q_2/\omega_1, \; Q_3/\omega_1\}$. Consequently, we can derive the value of γ that minimizes the summation Ψ using the algorithm **min_sum_df** shown in

Figure 3.6. The complexity of this algorithm is $O(N^2 L^2)$ assuming the summation on line 4 is $O(NL)$.

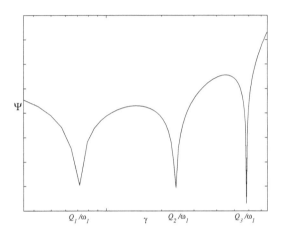

Figure 3.5: The effect of γ on Ψ for 3 $Q_i's$ and one ω_j.

$$(\gamma) = \mathbf{min\_sum\_df}\ (\{Q_1, ..., Q_N\}, \{\omega_1, \omega_2, ..., \omega_L\})$$

1. **Set** $\min\Psi = \infty$
2. **For** $j = 1$ to L
3.   **For** $i = 1$ to N
4.   **Set** $\gamma' = Q_i/\omega_j$ and $\Psi = \sum_{i=1}^{N} \sum_{j=1}^{L} U(Q_i, \gamma'\omega_j)$
5.   **If** $\Psi < \min\Psi$ **Then Set** $\gamma = \gamma'$
6.   **End For**
7. **End For**

Figure 3.6: Differentiation factor γ for minimum Ψ.

Theorem 3.1 also implies that determining the value of γ that minimizes $\psi(\pi_L^*)$ can be done by first substituting γ by one of the d_{ij} values, then obtaining the optimal

classification for each possible value of γ, and finally selecting the value of γ that has

the minimum $\psi(\pi_L^*)$. This is illustrated by algorithm **min_opt_df** in Figure 3.7. Note

that the optimal classification π_L^* may change by changing the value of γ. However, the

following theorem states that algorithm **min_opt_df** does indeed find the optimal value

of γ.

Theorem 3.2. *Algorithm min_opt_df finds the optimal value of*

$$\gamma^* \in \{d_{ij} = Q_i/\omega_j,\ i = 1, ..., N\ j = 1, ..., L\}$$

such that $\psi(\pi_L^*(\gamma^*), \gamma^*) \leq \psi(\pi_L^*(\gamma), \gamma)\quad \forall \gamma\ if\ U(Q_i, \gamma\omega_j)\ is\ a\ piecewise\ concave\ function,$

and $U(Q_i, \gamma\omega_j) = C_{ij} * U(Q_i/\omega_j, \gamma)$ *where* C_{ij} *is any constant.*

Proof. Given in Appendix A $\qquad\qquad\qquad\qquad\qquad\qquad\qquad\qquad\qquad\square$

By replacing line 6 of algorithm **min_sum_df** with a call to algorithm **OQC-FSL**,

we get the value of γ that minimizes $\psi(\pi_L^*)$ as shown in Figure 3.7.

From section 3.4, we know that the complexity of **OQC-FSL** is $O(N^2L)$. Therefore,

the complexity of this algorithm is $O(N^3L^2)$.

Example to illustrate the estimation of γ

Repeating the example in Section 3.2.3, with predetermined weights, and no capacity

restriction, we can have some flexibility on setting the value of the γ. The following cases

show the effect of the three different ways of estimating the value of γ on the value of

$\psi(\pi_L^*)$.

$(\gamma) = $ **min_opt_df** $(\{\Re\}, \{\omega_1, \omega_2, ..., \omega_L\})$

1. **Set** $\min\Psi = \infty$

2. **For** $l = 1$ to L

3.   **For** $i = 1$ to N

4.   **Set** $\gamma' = Q_i/\omega_j$

5. $\Psi = $ **OQC-FSL**$(\{\Re\}, (S_L))$

6.   **If** $\Psi < \min\Psi$

7.   **Set** $\gamma = \gamma'$

8.   **End If**

9.   **End For**

10. **End For**

Figure 3.7: Differentiation factor γ for minimum $\psi(\pi_L^*)$.

1. Using **median**(d_{ij}), γ=0.74222, $\psi(\pi_L^*)$=0.05070783, $\pi_L^* = \{[0.00033027, 0.0036963],$ $[0.024236], [0.043694, 0.074222]\}$.

2. Using algorithm **min_sum_df**, γ=0.24236, $\psi(\pi_L^*)$=0.07080461, $\pi_L^* = \{[0.00033027,$ $0.0036963, 0.024236], [0.043694], [0.074222]\}$.

3. Using algorithm **min_opt_df**, γ=0.74222, $\psi(\pi_L^*)$=0.05070783, $\pi_L^* = \{[0.00033027,$ $0.0036963], [0.024236], [0.043694, 0.074222]\}$.

Note that for cases 1 and 3, the solution is identical and the value of γ is the same while in both cases the value of $\psi(\pi_L^*)$ is lower than the value based on **min_sum_df**.

3.6 Service differentiation based on class weights, and limited link capacity

The link capacity C poses a restriction on the range of service differentiation values. We assume that the relationship between the capacity and the service differentiation factor is known through statistical modeling (e.g. such as the model described in Section 3.2.3) or through numerical estimation. From this relationship, we can obtain the minimum value of γ, or $\gamma_{\min}$ (Line 2 in Figure 3.8), which can then be used to modify the algorithm **min_opt_df** to obtain the optimal classification $\forall \gamma \geq \gamma_{\min}$. The modified algorithm is shown in Figure 3.8. The complexity of this algorithm is still $O(N^3 L^2)$.

The algorithm uses the result of Theorem 3.2 and selects a subset of possible values of $\gamma^* \in \{d_{ij} = Q_i / \omega_j, \ i = 1, ..., N \quad j = 1, ..., L\}$ which are greater than $\gamma_{\min}$ where $\gamma_{\min}$ is imposed by the limited link capacity.

3.7 Optimal classification for stochastic QoS

In practice, sometimes the set of QoS demands may not be explicitly known. Instead, the QoS requirements for the incoming traffic streams maybe determined by a continuous random variable Q, which has a probability density function (pdf) $f(Q)$ as shown in Figure 3.9. This QoS pdf may for instance be obtained empirically.

The optimization problem in this case will be to find the L set of ranges $\Re_L^* = \{R_1^*, R_2^*, ..., R_L^*\}$ defined on Q such that the expected value of the quantization overhead

$$(\psi(\pi_L^*), \pi_L^*) = \textbf{OQC-PCW} \ (\{\Re\}, \{\omega_1, ..., \omega_L\}, C)$$

1. Q_i/ω_j $\forall i \ \forall j$ are sorted in a non-decreasing order

2. **Initialize**: Set $\gamma_{min} = f(\{\Re\}, C)$, and $\min\Psi = \infty$

3. $\forall d_{ij} = Q_i/\omega_j$, $i = 1, ..., N \ j = 1, ..., L$ such that $d_{ij} \geq \gamma_{\min}$

4.   Set $\gamma = d_{ij}$, and $S_L = \{\gamma\omega_j, \ j = 1, ..., L\}$

5. $(\Psi, \Pi) = \textbf{OQC-FSL}(\{\Re\}, \{S_L\})$

6.   **If** $\Psi < \min\Psi$ **Then**

7.   **Set** $\gamma = \gamma$, $\ \psi(\pi_L^*) = \Psi$, and $\pi_L^* = \Pi$

8.   **End If**

Figure 3.8: OQC for predetermined class weights and limited link capacity (**OQC-PCW**).

$E(\psi(\Re_L))$ is minimized. In other words

$$E(\psi(\Re_L^*)) \leq E(\psi(\Re_L)) \quad \forall \Re_L$$

Where each of these ranges $R_i^* = \{Q_i^{\min}, Q_i^{\max}\}$ is defined by the minimum and maximum values of Q and $E(\psi(\Re_L))$ is defined as follows:

$$E(\psi(\Re_L)) = \sum_{l=1}^{L} \int_{R_l} U(Q, x_l) * f(Q) \ dQ$$

To leverage our solutions discussed before, we will sample the QoS pdf $f(Q)$ using a delta step ΔQ, such that $Q = k\Delta Q$, for $k = 1, ..., K$ where $K = Q_{\max}/\Delta Q$ is the sampling size. The $k-$th sample of $f(k\Delta Q)$ is scaled by ΔQ so that $\sum_{k=1}^{K} f(k\Delta Q) = 1$. The expected quantization overhead then becomes:

$$E(\psi(\Re_L)) = \sum_{l=1}^{L} \sum_{R_l} U(k\Delta Q, x_l) * f(k\Delta Q)$$

The system matrix A will change slightly such that $a_{kl} = U(k\Delta Q, x_l) * f(k\Delta Q)$ for k=1,..,K, and l=1,..,L. The **Dynamic_Program** algorithm can then be run against this modified matrix to obtain the ranges in $\Re_L^*$. Note that since $\lim_{K \to \infty} \Delta Q = 0$, the ranges $\Re_L^*$ obtained by this solution will approach the optimal ranges for large values of K.

3.8 Experimental Evaluation

In this section we examine the normalized quantization overhead defined as:

$$\psi_n(\pi_L^*) = \psi(\pi_L^*) / \sum_{i=1}^{N} Q_i$$

For the deterministic case, or

$$\psi_n(\Re_L^*) = E(\psi(\Re_L^*))/E(Q)$$

For the stochastic case.

We measure this overhead in different scenarios using the algorithms presented above. For the results presented in this section, we consider the link dropping probability as the prime QoS requirements. We assume, for example, the model explained in Section 3.2.3 when it is relevant and we use the utility function defined by Equation (3.1).

3.8.1 Results for deterministic QoS

Effect of increasing the number of service levels

We present first the results for *variable-sized* sets of connection requests ranging from N= 10 to 500 requests. For each set, the DP is uniformly distributed from 10^{-1} to 10^{-10}.

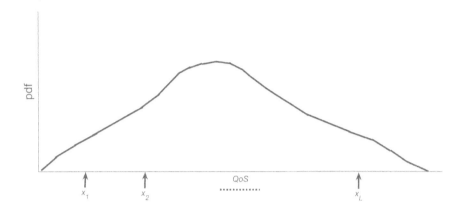

Figure 3.9: Example of a QoS probability density function.

The class weights are selected sequentially from the set $W = \{10^{-1}, 10^{-2}, ..., 10^{-8}\}$ such that for $L=2$, we use only the first 2 weights, for $L=3$, we use the first 3 and so on. We consider the case where the link is slightly overloaded (i.e. link load$\approx$1.1) and we use algorithm **OQC-FSL** to find an optimal classification based on fixed service levels calculated from the link's total load.

Figure 3.10 shows the results in this case. Each point in this figure is taken from the average of 10 sets of uniformly distributed DP values each of size N. We notice from the results that the increase in the normalized overhead is negligible when the number of service levels is above 3. This in fact should be an indication that the class weights are not selected efficiently. To further investigate this point, we show the results for $W = \{2^{-1}, 2^{-2}, ..., 2^{-8}\}$ as shown in Figure 3.11. We notice, in this case, that the change in overhead when increasing the service levels can reach 30% when the weights are selected

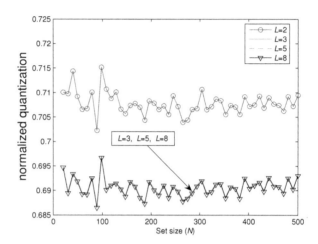

Figure 3.10: Variable sized sets for different number of service levels, $W = \{10^{-1}, 10^{-2}, .., 10^{-8}\}$.

appropriately. However, even in this case, it is clear that the gain of decreasing the normalized overhead is not significant when the number of service classes is above $L=5$. In fact, the gain in decreasing the overhead by increasing the number of service levels will become more significant as the link load increases. Figure 3.12 shows the impact of increasing the link load on the normalized overhead for a connection request size $N=500$ uniformly distributed. We see that the effect of increasing the number of service levels magnifies as the link becomes more overloaded. Interestingly, in this case, the effect does not appear to be significant for $L>5$ service levels.

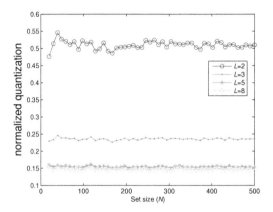

Figure 3.11: Variable sized sets for different number of service levels, $W = \{2^{-1}, 2^{-2}, .., 2^{-8}\}$.

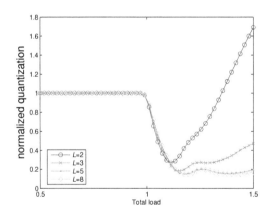

Figure 3.12: Variable load for different number of service levels, N=500, $W = \{2^{-1}, 2^{-2}, .., 2^{-8}\}$.

Effect of different estimation methods of γ on the normalized quantization

Figure 3.13 demonstrates the effect of different estimation methods of γ on the normalized quantization overhead when the QoS requirements have uniform distribution. The figure shows that the estimation of γ using the **min_sum_df** algorithm performs slightly better than the **median** in case of uniform distribution. The estimation using the **min_opt_df** performs consistently better than the 2 other methods.

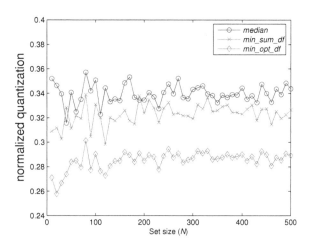

Figure 3.13: Variable sized sets for γ estimation methods, uniformly distributed QoS.

Effect of tolerance factor δ on the normalized quantization

Figure 3.14 and Figure 3.15 show the effect of the tolerance factor on the normalized quantization overhead. Figure 3.14 shows the tolerance effect for uniformly distributed

QoS values using the **OQC-FSL-ATF** algorithm for 4 service levels. We notice even when the link is lightly loaded, the quantization overhead is high. This is based on our original definition for the utility function. In other words, the link is best utilized when the load is close to the full load, in which case the quantization overhead is minimized. However, at certain loads, the link cannot offer the required QoS values, and in this case, some traffic streams might be rejected. This is illustrated by the discontinuation in the quantization overhead curve. We also notice that by increasing the tolerance, the link has more possibility of accommodating more traffic streams. Figure 3.15 shows the tolerance effect for uniformly distributed QoS values using the **OQC-PCW** algorithm using 4 service levels when the link is almost fully loaded. Clearly, increasing the tolerance factor plays a significant role in decreasing the quantization overhead regardless of the number of traffic streams. It is interesting to note that, increasing the tolerance factor limitations can be reduced through application-level adaptation. Our results are among the first to quantify the interrelation between application-level adaptation (tolerance) and QoS levels.

3.8.2   Results for stochastic QoS

In this section, we study the effect of the QoS statistical distribution on the value of the quantization overhead. For all results here we use $W = \{10^{-1}, 10^{-2}, ..., 10^{-8}\}$. Figures 3.16 to 3.19 show the results for the stochastic QoS requirements. Three important observations can be made from the six figures in addition to what was in the previous

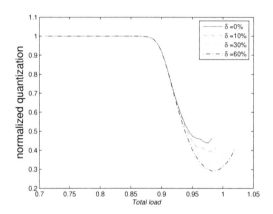

Figure 3.14: Effect of tolerance factor for different link loads, uniformly distributed QoS, using **OQC-FSL-ATF** algorithm.

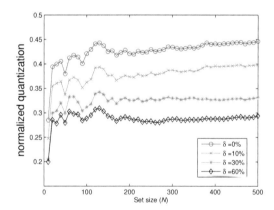

Figure 3.15: Effect of tolerance factor for variable-sized sets, uniformly distributed QoS, using the **OQC-PCW** algorithm.

section. First, the quantization overhead tends to converge to its optimal value as the value of K grows large and in fact this convergence is faster as we increase the number of service levels. This implies that the selected value of K depends on how many service levels used and the distribution of the incoming traffic streams. Apparently, we can use K as low as 100 (i.e. only 100 samples of the *pdf* function) for most of the QoS distributions for number of service levels more than 2. However, for low number of service levels (e.g. $L=2$ in Figure 3.17), it is recommended to use higher values for K. Second, the algorithm **min_sum_df** performs worse than the **median** for the clustered QoS distribution but still the **min_opt_df** outperforms both regardless of the QoS distribution. This indicates that the heuristic estimation for γ, while it is faster, its proximity to the optimal value is based on the input QoS distribution (see Figure 3.18). Finally, by increasing the tolerance factor progressively, the quantization overhead gain for the clustered distribution does not seem to increase with the same rate. This implies that by increasing the tolerance factor, not only may this become unacceptable by the applications generating traffic streams, but also the overhead gain may not be as significant.

3.9 Concluding Remarks

We have presented a group of algorithms for calculating the optimal classification for a set of traffic streams with diverse QoS requirements for a link model with a predetermined service levels or predetermined class weights. Our results show the effect of selecting the class weights based on the statistical distribution of the incoming connection requests. It

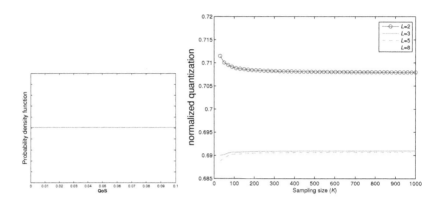

Figure 3.16: Effect of K for different number of service levels, uniform QoS distribution.

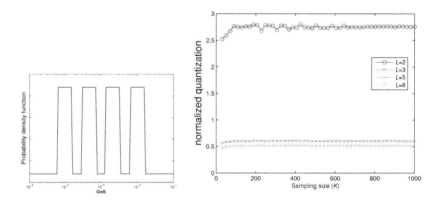

Figure 3.17: Effect of K for different number of service levels, for 4-cluster QoS distribution.

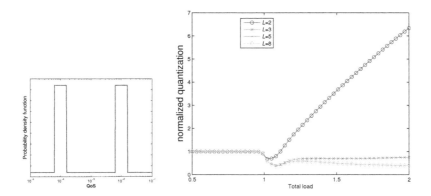

Figure 3.18: Effect of total load, for different number of service levels, for 2-cluster QoS

distribution for N=200.

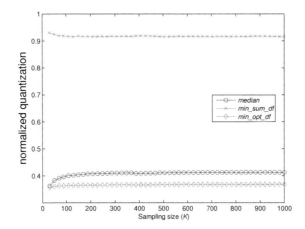

Figure 3.19: Effect of γ estimation methods for 2-cluster distribution using 5 service

levels.

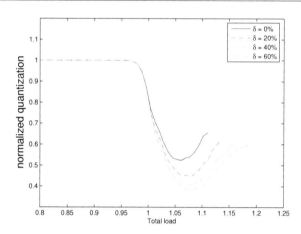

Figure 3.20: Effect of tolerance factor for different link load, for 2-cluster QoS distribution using 5 service levels.

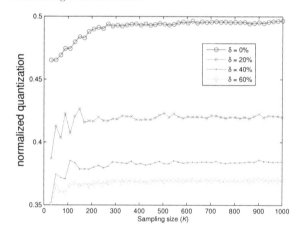

Figure 3.21: Effect of tolerance factor for variable-sized sets, for 4-cluster QoS distribution.

was shown that by carefully selecting the class weights, the quantization overhead can be significantly reduced especially for 4 or 5 service levels using dropping probability as the QoS metric. The results also show that the effect of increasing the number of service levels diminishes as the service levels rises more than 5. We also studied the effect of the three proposed γ estimation methods on the quantization overhead. The results show that the **min_opt_df** algorithm has the best performance over the other 2 heuristic methods regardless of the statistical distribution of the incoming QoS requirements. Furthermore, the expected quantization overhead for stochastic QoS requirements tends to converge as K (number of *pdf* samples) increases above 100 at least for all the QoS distributions discussed in this chapter. The exception from this is when we use service levels less than 3, in which case, K is recommended to be larger.

Chapter 4

QoS-Based Partitioning and Resource Allocation

In this chapter, we present two algorithms called **OQP** and **OQP-OBA** for calculating the optimal QoS-based partitioning and bandwidth allocation for a link model with a small set of variable service levels using dropping probability as the QoS metric. We formulate the partitioning process as a Dynamic Programming problem and present two polynomial time algorithms to obtain the optimal QoS-based partition with bandwidth allocation.

4.1 Introduction

We consider a link model with variable service levels which may be mapped to a finite number of MPLS Label-Switched-Paths (LSPs). Our target is to partition a set of traffic streams each with arbitrary local QoS-demand into a small number of classes and find the service level for each class such that the residual-allocated-resources as a result of

---

[0]A version of this chapter has been published [17]

the traffic partitioning is optimized. The residual allocated resources will be measured by the service quantization overhead which is the summation of the differences between the required QoS and the offered service level for all traffic streams. We formulate the partitioning process as a Dynamic Programming problem. We then present two polynomial time algorithms to obtain the QoS-based optimal partition with bandwidth allocation.

The problem of obtaining optimal quantized service levels for a set of connection requests has been investigated before in [61], [4], and [13]. Rouskas et al. in [61] defined a similar framework of assigning MPLS tunnels with specified data rates to a set of quantized service levels such that the performance penalty of wasted bandwidth is minimal. Both the deterministic and stochastic connection requests have been discussed. However, the case of having two joint requirements (e.g. bandwidth, and dropping probability) was not discussed. The authors in [4] discussed the problem in a different context where the set of requests are group of receivers on a multicast tree. Even though this context was different from our framework, the target was rather similar which is to find the small optimal set of offered rates (by the source) that maximizes the correlation between the total required rates requested by the receivers and the offered service levels. The solution for joint requirements was discussed in [13] for a simpler case where the set of service levels are predetermined. The study presented in this chapter extends these partitioning frameworks to the QoS based partitioning with resource allocation. It also offers solutions to the generalized problem where connections have both rate and QoS requirements and presents polynomial time algorithms for soft and hard QoS requirements.

The rest of the chapter is organized as follows; Section 4.2 formulates the model and the problem and relates our framework to QoS network architectures. The optimal solution for QoS-based partitioning using variable service levels is discussed in Section 4.3. The optimal partitioning with bandwidth allocation is outlined in Section 4.4. Section 4.5 depicts the results for applying our solutions to a set of QoS requirements with different distribution. Finally, Section 4.6 summarizes the key contributions in this chapter and highlights the importance of using optimal partitioning with bandwidth allocation.

4.2 Model and problem formulation

4.2.1 Model and notations

We consider a network of nodes that provide interconnection to a number of links. Each link has a *variable* set of L quantized service levels $S_L = \{s_1, s_2, ..., s_L\}$, such that each service level s_l, $l = 1, \ldots, L$ represents the QoS maintained in this class l. Each link also has a set of resources assigned to each class where some of these resources (e.g. link capacity C) might be shared amongst the service classes and are meant to be assigned arbitrarily to optimize certain criteria. Some other resources (e.g. buffer size B) might be assigned to the service levels permanently.

We also have a set of N connections $\Re = \{r_1, r_2, ..., r_N\}$. Each connection request r_i is defined using both bandwidth demand d_i and one local QoS requirement Q_i (e.g.

average delay, DP, jitter, etc.), namely $r_i \equiv \{d_i, Q_i\}$. If $r_i \in G_l$ and $s_l < Q_i$, this means

that traffic stream r_i is assigned to class l and is achieving better QoS than required and

that may entail a waste of link resources.

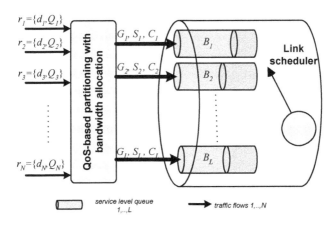

Figure 4.1: QoS-based partitioning model.

A partition $P_L = \{G_1, G_2, ..., G_L\}$ is defined for L service levels as a set of L groups

where each group G_l has one or more of the incoming traffic streams. Figure 4.1 shows

the QoS-based partitioning model including the link model with a logical queue assigned

for each service level and the resources assigned to each queue.

The best QoS for service class l is constrained by the resources and the total bandwidth

demands for the traffic streams assigned to this class. This relationship is characterized

by a deterministic function $f(\sum_{i \in G_l} d_i, C_l, B_l, ...)$ which is used to calculate the best achieved

QoS on a certain class given the allocated resources and the total load assigned to this class. This function can for example be obtained empirically or by using one of the appropriate well known Markov chain based models such as M/M/1/B [26, 67] for Poisson traffic. The authors in [41] and [68] also provide analysis and modeling for vast scheduling disciplines using Poisson and voice sources.

4.2.2 Optimization problem

The optimization problem is to find the optimal set of service levels $S_L^* = \{s_1^*, s_2^*, ..., s_L^*\}$, the optimal partition of traffic streams $P_L^* = \{G_1^*, G_2^*, ..., G_L^*\}$, and the optimal allocation of shared bandwidth $C_L^* = \{c_1^*, c_2^*, ..., c_L^*\}$ that minimize the QoS quantization overhead. The QoS quantization overhead is used to calculate the total penalty of supporting a small set of service levels on the granularity of providing QoS to different connections (flows) and is defined by the following objective function:

$$\psi(P_L) = \sum_{l=1}^{L} \sum_{\forall r_i \in G_l} U(Q_i, s_l) \tag{4.1}$$

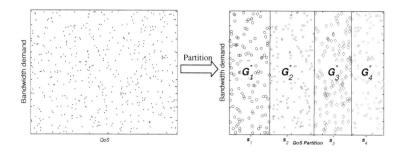

Figure 4.2: QoS-based partitioning.

$U(Q_i, s_l)$ is the utility function which measures the difference between the achieved and offered QoS. Figure 4.2 depicts the QoS-based partitioning process for a group of traffic streams defined by the 2 dimensional graph showing the bandwidth demand and QoS requirement for each traffic stream. It also shows the optimal service levels and the partitioned classes as a result of partitioning the traffic streams into 4 classes. The allocated bandwidth for each class (i.e. c_l^*) is selected such that $s_l^* \geq f(\sum\limits_{i \in G_l^*} d_i, C_l^*, B_l, ...)$. In other words, service level j cannot exceed the value $f(\sum\limits_{i \in G_l} d_i, C_l, B_l, ...)$ for a given amount of resources (i.e. c_l, B_l, etc.) allocated to this class. The optimization problem can then be formulated as follows:

$$\textbf{minimize} \quad \psi(P_L)$$

$$\textbf{subject to} \quad s_l \geq f(\sum_{i \in G_l^*} d_i, C_l, B_l, ...) \quad l = 1, ..., L \quad (4.2)$$

$$\sum_{l=1}^{L} C_l \leq C \quad (4.3)$$

Where C is the total link capacity shared amongst all service classes. In section 4.3 we will add another constraint to this optimization problem to force a maximum limit on the service level of each class which might be forced by the applications generating the traffic streams.

Without loss of generality, we consider a scheme of QoS-based partitioning where all traffic streams are sorted based on their QoS requirements and the utility function has certain constraints. The following definitions describe this partitioning scheme.

Ordered Partition

A partition $P_L = \{G_1, G_2, ..., G_L\}$ is called an ordered partition if $\forall r_i \in G_a$ and $\forall r_j \in G_b$, we have $Q_i \leq Q_j$, when $s_a < s_b$.

Utility Property

Intuitively, the utility function is non-increasing in the interval $Q_i \in [0, s_l]$ and it is non-decreasing in the interval $Q_i \in [s_l, \infty[$.

The importance of the these two definitions lies in the fact that if we choose a utility function that fulfills the utility property we only need to consider the ordered partitions when we search for an optimal solution (see Lemma 3.1). Indeed, considering only the ordered partitions will limit the number of possible solutions for optimality. However, in Chapter 3, we proved that brute force enumeration technique using ordered partitioning policy will lead to an exponential running time solution (see Lemma 3.2).

Utility functions

Selecting the appropriate utility function identifies the criteria used to minimize the objective defined by Equation (4.1). To illustrate this point we mention the following 3 cases:

1. We can select $U(Q_i, s_l)$ to measure the penalty of allocating unutilized resources for each service level by decreasing the service level s_l below the QoS values Q_i assigned to this class.

$$\text{Example}: \quad U(Q_i, s_l) = \begin{cases} \log(1 + Q_i - s_l) & s_l < Q_i \\ \\ 0 & \text{Otherwise} \end{cases} \quad (4.4)$$

2. We can instead select it to measure the penalty of violating the QoS Q_i required by each traffic stream.

$$\text{Example}: \quad U(Q_i, s_l) = \begin{cases} s_l - Q_i & Q_i < s_l \\ \\ 0 & \text{Otherwise} \end{cases} \tag{4.5}$$

3. Or we can measure the penalty in both cases to obtain the best set of service levels that achieves the trade-off between the allocated resources without violating the QoS requirements for the traffic streams.

$$\text{Example}: \quad U(Q_i, s_l) = |Q_i - s_l| \tag{4.6}$$

Hence, the support for all these aforementioned cases demonstrates the versatility of the proposed solutions discussed in this chapter. Notice that all the utility functions defined by Equations (4.4), (4.5), and (4.6) fulfill the utility property.

4.3 Optimal solution for QoS-based partitioning

This section describes the solution for the optimal QoS-based partitioning (OQP) problem without taking into consideration the allocation of the shared bandwidth. In other words, we will assume a simplified version of the problem where the traffic streams have only QoS requirements, and we need to partition them so that the quantization overhead is minimized. Even though, in this case we do not have to consider the set of constraints defined by Equations (4.2) and (4.3), which makes the problem easier. The following lemma states that the OQP problem is NP-Complete even for an arbitrary partition.

Lemma 4.1. *If* $U(Q_i, s_l)$ *is a step function defined as follows:*

$$U(Q_i, s_l) = \begin{cases} a_i & Q_i < s_l \\ \\ b_i & Otherwise \end{cases} \tag{4.7}$$

where a_{il} *and* b_{il} *are constants, then solving OQP for one specific partition* P_L *is equivalent to solving the SUBSET SUM problem (see SP13 in [69]) which is an NP-Complete problem.*

Proof. Given in Appendix A $\qquad\qquad\qquad\qquad\qquad\qquad\qquad\qquad\qquad\qquad$ □

In order to find the exact minimum value for OQP, we will impose another restriction on the utility function. Particularly, the following definition describes a specific class of utility functions for which we will develop the solution for OQP.

Piecewise concave functions

$U(Q_i, s_l)$ is a piecewise concave function if it is non-increasing concave in the interval $Q_i \in [0, \ s_l]$ and non-decreasing concave in the interval $Q_i \in [s_l, \infty[$. This means in addition to the fact that it fulfills the utility property, it is also concave on each interval separately. By concave function we mean that for any interval $Q_i \in [a, \ b]$, $U(Q_i)$ satisfies this inequality

$$U(\alpha\, a + (1-\alpha)b) \quad \geq \alpha\, U(a) + (1-\alpha)\, U(b) \quad 0 \leq \alpha \leq 1 \tag{4.8}$$

Figure 4.3 shows an example of a piecewise concave function which is defined as follows:

$$U(Q_i, s_l) = \log(1 + |Q_i - s_l|)$$

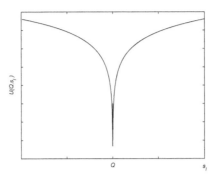

Figure 4.3: Example of a piecewise concave function.

The following theorem then states the rule for finding the optimal service levels to mini-

mize the quantization overhead.

Theorem 4.1. *If all $U(Q_i, s_l)$ $\forall Q_i$ are piecewise concave functions, then for each group*

G_l^* in the optimal partition P_L^*, the value of the service level s_l^* that coincides with one of

the values $Q_i \in G_l$ will always minimize the total overhead of class j such that $U(G_l, s_l^*) \leq$

$U(G_l, s_l)$ $\forall s_l \notin \{Q_i : i \in G_l\}$.

Proof. Given in Appendix A   □

This means that for each class $G_l \in P_L$ the service level s_l must be one of the values

in the set $\{Q_i: i \in G_l\}$ in order to minimize the summation $U(G_l, s_l^*) = \sum_{\forall Q_i \in G_l} U(Q_i, s_l^*)$.

One of the most efficient ways of solving this problem is by recursive enumeration which

is the main idea of dynamic programming. Hence, the problem can be mapped to the following dynamic program.

$$\psi(i,l) = \min_{1<k<i-1}[\psi(k,l-1) + \min_{k+1<m<i}\sum_{j=k+1}^{i}U(Q_j,Q_m)] \tag{4.9}$$

where $\psi(i,l)$ is the local optimal quantization overhead for requests up to request i using l service levels. This local optimal value is calculated based on recursive enumeration. Finding the value of k that minimizes Equation (4.9) is used to find the current best partition class $G_l^* \in P_L^*$. In other words $G_l^* \equiv \{Q_{k+1}, Q_i\}$, such that

$$k = \arg\min_{1<k<i-1}[\psi(k,l-1) + \min_{k+1<m<i}\sum_{j=k+1}^{i}U(Q_j,Q_m)]$$

Equation (4.9) utilizes the result of theorem 4.1 in calculating the minimum class quantization overhead as follows:

$$U(G_l \equiv \{Q_{k+1},...,Q_i\}, s_l^*) = \min_{k+1<m<i}\sum_{j=k+1}^{i}U(Q_j,Q_m)$$

$$\text{And}\qquad s_l^* = \arg\min_{Q_m\in\{Q_{k+1},...,Q_i\}}\sum_{j=k+1}^{j}U(Q_j,Q_m) \tag{4.10}$$

One thing to notice in Equation (4.10) is that the selected service level s_l^* that minimizes $U(G_l \equiv \{Q_i,...,Q_j\}, s_l^*)$ may not be accepted by the applications generating the traffic streams. This is because the difference between the service level s_l^* and the QoS requirements maybe more than what the applications can tolerate. To account for this, either we have to incorporate the tolerance factor of each traffic stream into the utility function or we can add the tolerance factor as an additional parameter and select the service level such that the following condition is satisfied:

$$((1+\delta)\ Q_i - Q_m)\ \geq 0$$

where δ is the maximum percentage of QoS drop that can be tolerated by any traffic stream r_i. If $\delta=0$, this means that the traffic stream cannot tolerate any service level drop. Hence, the minimum class quantization overhead (MCQO) can be redefined as follows:

$$U(G_l \equiv \{Q_{k+1}, ..., Q_i\}, s_l^*) = \min_{\substack{k+1<m<i \\ ((1+\delta) \, Q_j-Q_m) \, \geq 0}} \sum_{j=k+1}^{i} U(Q_j, Q_m) \qquad (4.11)$$

Based on this definition we can now write the dynamic program for solving OQP as shown in Figure 4.4.

Optimal QoS-based partitioning [OQP] $(\{Q_i\}_1^N, L)$

1. Q_i $\forall i$ are sorted in a non-decreasing order
2. **For** $l = 1$ to L
3.   **For** $i = 1$ to N
4. $\psi(i, l) = \min\{ \min_{1<k<i-1} [\psi(k, l-1) + U(\{Q_{k+1}, ..., Q_i\}, s_l^*)], \psi(i, l-1)\}$
5.   **End For**
6. **End For**
7. **Return** $\psi(N, L)$

Figure 4.4: Algorithm optimal QoS-based partitioning (OQP).

The complexity for lines 2 to 4 in Figure 4.4 is $O(N^2 L)$. The class utility defined by Equation (4.11), can be pre-computed for all $i=1, \ldots, N$, and $j=1,\ldots,N$ with a complexity of $O(N^3)$ (a similar pre-computation is explained in details in [4]). Therefore, the total complexity of **OQP** is $O(N^3)$.

4.4 Optimal QoS-based partitioning with optimal bandwidth allocation (OQP-OBA)

This section describes the solution for the optimal QoS-based partitioning with optimal bandwidth allocation (**OQP-OBA**) problem. In this case, we have a limited link capacity (as an example of a shared resource) that we divide freely amongst the QoS partition classes. This will limit the capacity assigned to each class, and hence, will affect the number of network flows that can be admitted to that class and the service level that this class can provide.

Assuming the allocated bandwidth vector is $C'_L = \{c'_1, c'_2, ..., c'_L\}$ (i.e. $\|C'_L\| = \sum_{i=1}^{L} c'_i \leq C$), then for each class, the service level s^*_l is constrained by c'_l through the following inequality

$$s^*_l \geq f(\sum_{i \in G^*_l} d_i, c'_l, B_l, ...)$$

In this case, the MCQO in Equation (4.11) will become

$$U(G_l \equiv \{Q_i, ..., Q_j\}, s^*_l, c'_l) = \min_{\substack{i < m < j \\ Q_m \geq f(\sum_{i \in G^*_l} d_i, c'_l, B_l, ...)}} \sum_{k=i}^{j} U(Q_k, Q_m) \qquad (4.12)$$

The problem in this case becomes NP-Hard, and therefore, in order to get the optimal bandwidth allocation C^*_Lm, we will divide the capacity C into small parts using Δc where Δc is small. The bandwidth assigned to each class will then be $w \cdot \Delta c$, where $w = 1, ..., W$ and $W = \lfloor C/\Delta c \rfloor$ is the number of capacity divisions. The recursive equation for the dynamic program will then become.

$$\psi(i,l,w) = \min_{\substack{1<k<i-1\\1<b<w-1}} [\psi(k,l-1,b) + \min_{\substack{k+1<m<i\\\lambda=\Delta c(w-b)}} U(\{G\}_{k+1}^i, Q_m, \lambda)] \tag{4.13}$$

Figure 4.5 shows the **OQP-OBA** dynamic program. Similar to the OQP case, we find

the value k that minimizes Equation (4.13) for a particular bandwidth allocation b as

follows:

$$[k,b] = \arg\min_{\substack{1<k<i-1\\1<b<w-1}} [\psi(k,l-1,b) + \min_{\substack{k+1<m<i\\\lambda=\Delta c(w-b)}} U(\{G\}_{k+1}^i, Q_m, \lambda)]$$

Then we get c_l^* as follows:

$c_l^* = \Delta c(w-b)$ and s_l^* as follows:

$$[s_l^*] = \arg\min_{Q_m \in \{Q_{k+1}, \dots, Q_i\}} U(\{G\}_{k+1}^i, Q_m, c_l^*)$$

OQP with Optimal Bandwidth Allocation [OQP-OBA] $(\{Q_i\}_1^N, L)$

1. $Q_i \quad \forall i$ are sorted in a non-decreasing order

2. **For** $l = 1$ to L

3.   **For** $i = 1$ to N

4.   **For** $w = 1$ to W

5. $\psi(i,l,w) = \min_{\substack{1<k<i-1\\1<b<w-1}} [\psi(k,l-1,b) + \min_{\substack{k+1<m<i\\\lambda=\Delta c(w-b)}} U(\{G\}_{k+1}^i, Q_m, \lambda)]$

6.   **End For**

7.   **End For**

8. **End For**

9. **Return** $\psi(N, L, W)$

Figure 4.5: Algorithm OQP with optimal bandwidth allocation (**OQP-OBA**).

The complexity for lines 2 to 8 in Figure 4.5 is $O(N^2W^2L)$. The class utility defined

by Equation (4.12), can be pre-computed for all $i=1,\dots, N$, $j=1,\dots,N$, and $w =$

1,...,W with a complexity of $O(N^3W)$. So, the total complexity of **OQP-OBA** is $O(\max\{N^2W^2L, N^3W\})$.

4.5   Experimental evaluation

In order to verify the feasibility of our algorithms, we consider other schemes for partitioning that we can use to measure the advantage of using **OQP-OBA**. First, we consider the trivial arbitrary QoS-based partitioning with arbitrary bandwidth allocation (AQP-ABA). We also consider two other cases where either partitioning or bandwidth allocation is done arbitrarily and the other is using optimal value (i.e. **OQP-ABA**, and **AQP-OBA**). For all these cases, we measure the normalized quantization overhead as follows:

$$\psi_n(P_L^*) = \psi(P_L^*)/\sum_{i=1}^{N} Q_i$$

We take the link's DP with different distributions in the range from 10^{-1} to 10^{-10} as an example for QoS. We measure the normalized quantization for the different cases mentioned above and compare it with the **OQP-OBA** algorithm discussed in section 4.4.

Figure 4.6 depicts the difference between Arbitrary Partitioning (AP) and Optimal Partitioning (OP) for a link with 5 service levels and QoS requirements distributed in 5 clusters and uniform bandwidth requirements as shown in the figure. We notice when using AP that the groups might be erratic and some flows may end up being assigned a much higher/lower QoS than what they ask for which causes discrepancy in utilizing the link resources. OP, on the other hand, guarantees proper logical grouping for the flows

and hence provide consistency in utilizing the link resources.

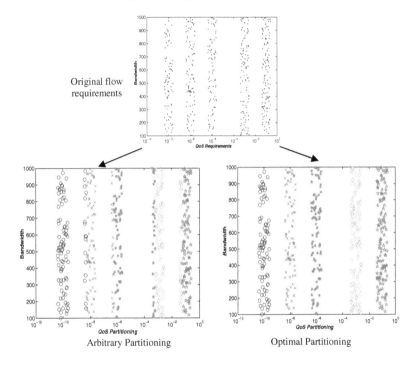

Figure 4.6: Difference between arbitrary and optimal partitioning.

Figure 4.7 shows the normalized quantization overhead against variable sized sets

of flows for a link with 5 service levels, QoS requirements distributed in 5 clusters and

uniform bandwidth demands. We notice that **OQP-OBA** consistently outperforms the

other schemes by at least 50% for clustered QoS distribution. This is because **OQP-**

OBA guarantees the best logical grouping for the flows and selects the service level that best fits each group. We also notice that **OQP-ABA** outperforms the other 2 schemes which indicates that the effect of logical grouping is more crucial than allocating the bandwidth for clustered QoS distribution.

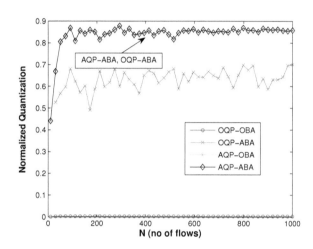

Figure 4.7: Variable sized sets for 5 service levels.

Figure 4.8 depicts the effect of the link's total load on the normalized quantization for different partitioning and bandwidth allocation schemes using 5 service levels, 5-cluster QoS distribution and uniform bandwidth demands. Again, we notice that **OQP-OBA** consistently outperforms the other schemes even for low link loads. We also notice that after specific link load, which is different for each case, the link cannot provide the QoS required without violating the QoS requirements on each group significantly. Therefore, even if the link is overloaded, **OQP-OBA** guarantees graceful service degradation for high link loads. This is a crucial feature especially in the absence of admission control mechanism.

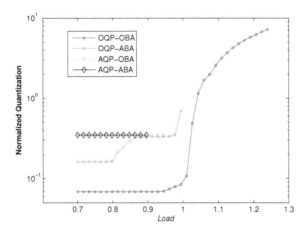

Figure 4.8: Effect of link load for different partitioning and bandwidth allocation schemes.

Figure 4.9 shows the effect of increasing the number of service levels on the normalized quantization using difference QoS distributions. We notice here that beyond 4 or 5 levels the normalized quantization tends to be almost constant regardless of the QoS distribution used. Since increasing the number of service levels may affect the complexity of the algorithm, this crucial result indicates that using 4 or 5 service levels will achieve the trade-off between complexity and minimizing the normalized quantization overhead.

Finally, Figure 4.10 shows the effect of increasing the capacity divisions on the convergence of the **OQP-OBA** algorithm for different number of service levels using uniform

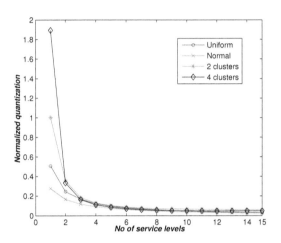

Figure 4.9: Number of service levels for different QoS distributions.

distribution for QoS requirements and bandwidth demands. We notice that the convergence happens for all values of L but tends to be faster for low values (e.g. $L=2$). However, even for high values of L, the convergence happens for a value of W less than 50.

4.6   Concluding Remarks

In this chapter, we have presented 2 algorithms for calculating the optimal partitioning and bandwidth allocation for a set of traffic streams with diverse QoS requirements for a link model with variable service levels. Such algorithms can be useful for service providers to design the network service levels that achieve the best granularity based

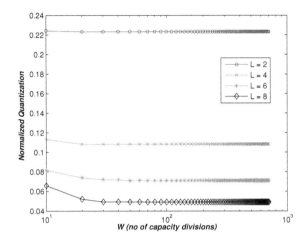

Figure 4.10: Number of service levels for different QoS distributions.

on the different distributions for the QoS requirements and bandwidth demands. Our results using dropping probability as the QoS metric show that using 4 or 5 service levels will achieve the trade-off between complexity and granularity irrespective of the QoS distributions. They also show that using **OQP-OBA** guarantees the best logical grouping and best selected service levels which achieve the best granularity. Moreover, using **OQP-OBA** will guarantee graceful service degradation when the link is overloaded in the absence of admission control.

Chapter 5

Optimal Resource Allocation for

Homogeneous Wireless Multicast

In this chapter, we present a novel *decentralized* algorithm called **ORAWM** that achieves the optimal rates for a set of homogeneous multicast sessions such that the aggregate utility for all sessions is maximized. We present the formulation of the multicast resource allocation problem as a non-linear optimization model and highlight the cross-layer framework that can realize this problem in a real distributed ad hoc network environment with asynchronous computations. We also present a series of implementations based on different network settings and show that not only convergence to the optimal rates is attained in all these network settings but also network changing conditions such as mobility and dynamic channel capacity can be tracked in real time.

---

[0]A version of this chapter has been published [18, 19, 20]

5.1 Introduction

The rapid growth of customer demands for fast, group-oriented, and mobile wireless services has mandated the need for inventing new wireless-based techniques that are scalable, bandwidth-efficient, and simple to implement and work with the existing wireless standards. Ad hoc networks empowered by multicast-based communication scheme is one of the potential strategies to achieve these features. Wireless ad hoc networks support peer-to-peer communication between active nodes through wireless links, and therefore, the network topology changes by changing the location of the wireless nodes caused by mobility. Active node has routing capability that allows it to create multihop path to any other node and, depending on its location, can forward packets for its neighboring nodes through that path. Therefore, ad hoc networks are extremely flexible, self-configurable, and do not require any infrastructure for their operation. This has opened the door for the possibility of providing many applications over wireless such as Videoconferencing, TV-over-wireless, intelligent transportation/traffic systems, field games, rescue and disaster recovery, and military operations. This consequently has led to the potential of using multicast as a bandwidth-efficient technique for communication.

The one hop broadcast characteristic of the MAC layer in wireless ad hoc networks has triggered the use of multicast communication scheme as one of the natural strategies that can multiply the overall network throughput with very limited overhead. This is because multicast packets are forwarded once to reach all the multicast members in the neighborhood using a single transmission, and this effect increases even more in multi-hop ad hoc networks.

Multicast, however, introduces new challenges due to its complexity, and lack of standards and techniques that can effectively allocate network resources. Efficient resource allocation is required to guarantee optimal resource utilization while providing system-wide fairness to different network flows end-to-end. The challenge in multicast resource allocation is that, not only the resource allocation has to optimize access to network resources for each source-destination pair, but also it has to relate different source-destination pairs on the same multicast group to try to utilize the bandwidth efficiency of each multicast session (i.e intra-flow bandwidth efficiency) and achieve system-wide fairness across all multicast and unicast sessions (i.e. inter-flow fairness).

One fundamental difference between multihop wireless networks and wired networks, in addition to the broadcast nature of the MAC layer, is the concept of location-based contention. In multihop wireless networks, flows may contend for the channel as long as they are within the interference range of each other, even if they are transferred on separate virtual wireless links. This imposes the problem of forming contention domains as part of resource allocation in order to map the logical network topology into the physical channel characteristics. The problem becomes trickier in case of multicast where different branches of the same multicast group may incur different levels of contention based on the location of the multicast members.

In this chapter, we present an optimization model that captures the problem of multicast resource allocation in wireless ad hoc networks and discuss the different possible solutions to that model based on the network changing conditions and information distribution amongst its wireless nodes. This optimization model must guarantee steering the *entire* network towards the optimal point in real time, and hence react to network conditions as they occur. Also, mapping the multicast resource allocation problem into an optimization model has the advantage of leveraging many efficient optimization techniques to devise and improve the problem solutions.

The problem of resource allocation for unicast flows has been investigated before in [1, 10], and [6]. A common pricing mechanism has been used in all of them where each network resource calculates a *price* that represents the relationship between the load on this network resource and the capacity that it can offer. A *utility* is associated with each end-to-end flow to represent the flow's resource requirement. The objective then is to maximize the aggregate utility function for all flows subject to network resource constraints to achieve optimal resource utilization while providing system-wide fairness across all flows. The authors in [10] and [1] have studied the resource allocation in wired networks where they used the pricing mechanism to calculate a price on each link and then send this price through feedback to sources to calculate the required rate that optimizes the aggregate utilities of all network sources. Yuan et al. in [6] considered a similar pricing mechanism for allocating end-to-end unicast flow rates in ad hoc networks. A mechanism has been devised there to model the wireless nodes based on their locations into contention domains and then use them as the base for calculating the load on each contention domain versus the capacity offered by the wireless channel. None of these studies, however, considered the multicast flows where resource allocation mechanism relates the multicast source-destination pairs to benefit from the bandwidth efficiency of multicast as a communication scheme (see example in Section 5.2.2). Resource allocation for multicast wired networks has been studied in [44, 45] and [7]. They define the multicast tree between one source and multiple receivers as group of terminal (non-junction) nodes and branch (junction) nodes and each downstream on a branch node contribute to the load of the link that the downstream traverses. However, as we will see in the following sections, the location based contention, one hop broadcast of the wireless MAC layer, and the absence of any centralized control in the multihop wireless networks make the problem in our case much more complex.

The remainder of this chapter is organized as follows. In Section 5.2, we formulate the problem and provide some terminology. Our solution approach is presented in Section 5.3. We present our algorithm with its improvement to work in asynchronous network environments in Section 5.4. We then discuss the implementations of our algorithm in different network settings and provide a simulation-based study in Section 5.5. Finally, we conclude in Section 5.6.

5.2 Model and problem formulation

We describe the model and explain the notations used in this chapter. Then we present a mathematical formulation of the optimization problem.

5.2.1 Model and Notations

We consider a wireless ad hoc network consisting of a set of nodes V spread over a wireless space each with a specific transmission range and interference range. We exploit the protocol model explained in [70] and leveraged in [6] for wireless packet transmission. In this model, the transmission from node i is successfully received by node j $(i, j \in V)$ if (1) the distance between the two nodes is no more than a certain range (i.e. transmission range), and (2) for every other node $k \in V$ simultaneously transmitting over the same channel, the distance between j and k is more than a specific range (i.e. interference range). For some protocols which require acknowledgment from j to i (e.g. IEEE 802.11 MAC), node i is also required to be interference free at the time of sending the acknowledgment.

We model the wireless ad hoc network as a *directed* graph $G = (V, E)$ where E is the set of wireless virtual links produced as a result of nodes located within the transmission range of each other. Each wireless link $e \in E$ has two end nodes i, and j (i.e. $e = \{i, j\}$).

The network is shared by a set of M *end-to-end* multicast groups. Each multicast group m has a unique source s_m, a set of receivers $R_m = \{r_{m1}, r_{m2}, ...\}$, and uses a set of wireless links E_m and a set of nodes V_m for either receiving or relaying traffic. Thus, a multicast group $m \in M$ is defined by the triplet $\{s_m, R_m, E_m\}$. Each multicast group has a rate x_m which is allowed to vary in the interval $I_m = [w_m, W_m]$ [1], and I is the set of all such intervals. A fundamental difference between the unicast and multicast case, is the fact that one hop broadcast may be used to transfer traffic from one source to one or more destinations. To capture this notion, the one hop data transmission from one node

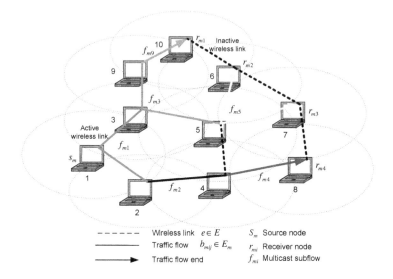

Figure 5.1: Wireless ad hoc network model.

i to a set of nodes $J \subseteq V_m$ within the multicast flow m along *one* or *more* wireless links

(branches) is referred to as a multicast subflow of m or f_{mi}. The set of multicast subflows

on a multicast group m is referred to as $F_m = \{f_{m1}, f_{m2}, ...\}$, as shown in Figure 5.1.

The set of all multicast subflows for all the M groups is referred to as F. Each multicast

subflow uses one or more wireless links from one source i to a set of destinations $J \subseteq V_m$,

i.e. $f_{mi} = \{b_{mij} : \forall b_{mij} = \{i, j\} \ j \in J\}$, with a cardinality K_{mi} equal to the number of

branches of f_{mi}. We also define an active wireless link $a_{ij} \in E$ to be the wireless link

that carries traffic from at least one multicast group, and $A \subseteq E$ is set of all such links.

a_{iJ} refers the *aggregated* multicast subflow which is represented by the set of active links

$a_{ij} \ j \in J$ that are used by one or more multicast subflows $f_{mi} \ m \in M$, simultaneously.

Based on the protocol model (explained in [48]) the traffic from two different subflows on two active wireless links contend with each other if either the source or the destination of one wireless link is within the interference range of the source or destination of the other wireless link. This means, that only one node on these two contending links may transmit packet at any given time. It is important to notice that for a multicast subflow, different branches may experience different levels of contention based on the location of each destination on the subflow. For example, in Figure 5.1, although branches b_{m57}, and b_{m56} belong to the same multicast subflow, they suffer different contention. The traffic on b_{m57} contends with the set of active wireless links $\{(a_{35}, a_{39}), a_{48}, a_{24}, a_{13}\}$, whereas the traffic on b_{m56} contends with the set of active wireless links $\{(a_{35}, a_{39}), a_{48}, a_{24}, a_{13}, a_{910}\}$ where links grouped with brackets identify the subgroups of links that carry the traffic from the same subflow. This means that b_{m57} experiences less contention than b_{m56}, and the same result could be derived for branches b_{m12}, and b_{m13}. Notice that a_{56} and a_{57} do not contend because they carry the traffic from the same subflow[1]. Due to the broadcast nature of the MAC layer and based on the scheduling mechanism used, if we increase the rate on these subflows beyond certain limit, some branches may suffer low throughput [72]. This intraflow contention diversity maybe common in case of asymmetric network topology like the case of Figure 5.1.

For simplicity, we assume that a packet is successfully transmitted over a multicast

---

[1]Here we assume that the multicast scheduler will guarantee that the acknowledgments from the group receivers (if any) may never contend. This can be achieved easily by assuming that each receiver has a certain time slot to send the acknowledgment back to the source [71].

subflow f_{mi} if (1) the packet reaches all destination nodes J on all the branches b_{mij}; and (2) acknowledgments (using the notation of IEEE 802.11 MAC standards) have been received successfully from all these destinations back to the source node i [71]. This means that the maximum rate for subflow f_{mi} cannot exceed the rate that can be withstood on the most contended branch of that subflow. Based on this assumption, the protocol model can be extended for multicast subflows as follow; the traffic from two different subflows on a group of active wireless links contend if either the source or *any* of the destinations of one subflow is within the interference range of the source or *any* of the destinations of the other subflow.

To model the contention between the active wireless links, we use a contention domain mechanism [48] by forming a logical contention graph $G_c = (V_c, E_c)$. Each vertex on that graph corresponds to the aggregated multicast subflow a_{iJ} which carries the traffic from *one* or *more* subflows simultaneously. The links between two vertices represent whether the traffic on the two aggregated subflows contend with each other, which means *any* of the active links of one aggregated subflow is located in the contention region of *any* of the active links of the other aggregated subflow. A complete subgraph in G_c is referred to as a *clique*, and a *maximal clique* is the clique that is not part of any other cliques. This maximal clique represents the maximal set of active wireless links that contend with each other, which means that one "subflow" within this clique may transmit a packet at any given time [6]. Therefore, the sum of the rates over this maximal clique cannot exceed the channel capacity achieved by using a particular scheduling mechanism in the MAC layer (e.g. IEEE 802.11 DCF). Similar to the *link* in wired networks, the maximal clique is the resource entity that is used to drive the constraints in wireless ad hoc networks. As we will see later, a special consideration should be given to multicast subflows that have more than one branch in the same maximal clique in deriving these constraints. The following inequality formulates these set of constraints:

$$\sum_{m:(F_m \cap V_c^q) \neq \emptyset} x_m \leq c_q \qquad \forall q \in Q \tag{5.1}$$

where q is a maximal clique in the set of all maximal cliques Q, c_q is the achieved channel capacity for clique q based on the scheduling discipline used in the MAC layer, and $V_c^q \subseteq A$ are the set of vertices in G_c that belong to clique q. Note that unlike the unicast case [6] (where the active wireless link is used as entity to calculate the upper bound), we use the multicast subflow (which groups one or more branches together) as the entity for calculating the upper bound within any maximal clique. This will reduce the number of constraints by removing any redundancy, while keeping the constraints linear and that will simplify the solution for the resource allocation problem as we will see later.We say that a multicast subflow $f_{mi} \in F$ uses the resources in a maximal clique q if at least one branch of f_{mi} uses an active wireless link that exists as a node on V_c^q (see implementations in Section 5.5). In practice, the channel capacity achieved by a specific MAC scheduler may vary even with time. Thus, not only does the resource allocation algorithm have to converge quickly to the optimal values for multicast rates, but also, through online calculations, it has to track the changes in the achieved channel capacity of each maximal clique to maintain the trade-off between resource utilization, and system-wide fairness.

5.2.2   Mathematical Formulation

Each multicast group $m \in M$ is associated with a *utility function* $U_m(x_m)$ which measures the degree of satisfaction based on assigning a specific rate x_m. The optimization problem then is to find the set of rates assigned to all multicast groups in M such that the aggregated utility function [2] of all multicast groups, commonly referred to as the *social welfare*, is maximized. This can be formulated as a non-linear optimization problem as follows:

$$\mathbf{P}: \quad \textbf{maximize} \quad \sum_{m \in M} U_m(x_m) \tag{5.2}$$

$$\textbf{subject to} \quad \Gamma x \leq C \quad \forall q \in Q \tag{5.3}$$

$$x_m \in I_m \quad \forall m \in M \tag{5.4}$$

Equation (5.3) is the matrix form of (5.1) where x is a vector of all multicast group rates, and C is the vector of achieved capacities on all maximal cliques. Γ is the *clique-group* matrix with each element in Γ represents the number of multicast subflows within a maximal clique q. Throughout the rest of the chapter, we will make the following assumptions to facilitate the solution for the *primal* problem $\mathbf{P}$:

Assumption 1: There exists at least one vector $\tilde{x} \in I$ such that $\sum_{m \in (F \cap V_c^q)} \tilde{x}_m \leq c_q \quad \forall q \in Q$ and $\tilde{x}_m \in I_m$ (i.e. $\sum_{m \in (F \cap V_c^q)} \tilde{w}_m \leq c_q \ \forall q \in Q$). This means that admission control is out of the scope of this model.

---

[2]We assume here that the utility function is additive, which means that the aggregated utility function for allocated vector x is the summation of all utility functions as defined by Equation 5.2.

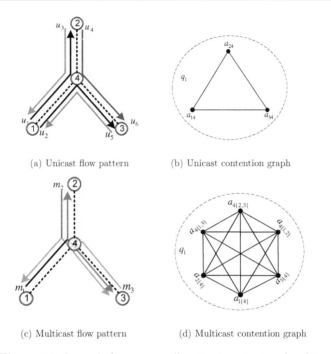

(a) Unicast flow pattern   (b) Unicast contention graph

(c) Multicast flow pattern   (d) Multicast contention graph

Figure 5.2: Example for resource allocation in unicast and multicast.

Assumption 2: On the interval I_m, all the utility functions $U_m \ \forall m \in M$ are increasing, strictly concave and twice continuously differentiable. Thus $-U_m'' \geq \gamma_m > 0$.

The importance of these 2 assumptions lies in the fact that Assumption 1 implies that problem $\mathbf{P}$ is feasible , i.e. it has a solution, while Assumption 2 forces this solution to be unique.

Next, we present an example to illustrate the notations and highlight the difference between unicast and multicast in allocating rates in an ad hoc network. Figure 5.2 shows an example of an ad hoc network where there are 4 nodes: 1, 2, 3, and 4 connected through wireless links as shown in the figure. Each of the Nodes 1, 2, and 3 needs to send traffic to the other 2 nodes, and Node 4 acts as the relay node for all traffic going from one node to another. Figure 5.2(a) shows the flow pattern to implement this scenario using only unicast flows. We notice here that each of the Nodes 1, 2, and 3 must send 2 unicast flows to the other 2 nodes . As shown in the figure, we need 6 unicast flows $\{u_1, ..., u_6\}$, with rates represented by the vector $x' = \{x_1, ..., x_6\}$. The aggregated subflows on each wireless link is represented by one node in the contention graph as shown in Figure 5.2(b). We notice here that all the subflows on all the links mutually contend with each other forming one maximal clique with an achieved capacity c_1' as shown in Figure.

On the other hand, Figure 5.2(c) shows the flow pattern in case of using multicast flows. We notice here that each node uses only one multicast group to send traffic to the other two nodes. We notice also that the multicast subflows(e.g. the one going from Node 4, to nodes 1, and 2), can leverage the broadcast nature of the MAC layer by using single packet broadcast to send data from Node 4 to the other two nodes simultaneously. Using this fact to implement this scenario, we only need 3 multicast groups $\{m_1, m_2, m_3\}$ for which rates are represented by a vector $x = \{x_1, x_2, x_3\}$. Figure 5.2(d) shows the resulting contention graph representing each group of links used by multicast subflow as a node on the graph. Here too the graph shows that all the subflows form one maximal clique q_1, which means that only one node in the network can send a packet at any given time. Assume the achieved capacity on this clique is c_1, then we can derive the constraints in Equation (5.3) for unicast and multicast as follows:

$$\textbf{unicast}: \quad \begin{pmatrix} 2 & 2 & 2 & 2 & 2 & 2 \end{pmatrix} x' \le c_1'$$

$$\textbf{multicast}: \quad \begin{pmatrix} 2 & 2 & 2 \end{pmatrix} x \le c_1$$

This shows that by using the notion of multicast subflow, which groups multiple branches together, the clique-group matrix dimension is reduced significantly in case of multicast. This is a key feature for multicast resource allocation especially for large networks. In this simple and symmetric topology, the solution can be guessed intuitively. Hence, to achieve optimum resource utilization while maintaining fairness amongst all groups, the elements in each of the vectors x' and x must all have same value and the optimal value for each case can be calculated as follows:

$$\textbf{unicast}: \quad x_u' = c_1'/12 \quad u = 1, \cdots, 6$$

$$\textbf{multicast}: \quad x_m = c_1/6 \quad m = 1, 2, 3$$

which means that by using multicast, not only the number of flows is reduced (which means the processing overhead on each node is reduced), but also, the optimum allocated rate for each multicast group is close to double the value of the optimum allocated rate on any unicast flow (assuming the achieved capacities are almost equal, i.e. $c_1 \approx c_1'$). It is not hard to recognize that the effect of multicast intra-flow bandwidth efficiency will become even stronger when we increase the number of hops.

5.3 Solution Approach

Solving the resource allocation problem **P** centrally would require the knowledge of the utility functions and the knowledge of all contention domains and multicast groups which is impractical. Instead, we propose a decentralized scheme that minimizes the coordination between networks nodes and adapts naturally to network changes. The key to this is the use of the duality theory [73] which suggests solving the dual problem by introducing additional dual variables called *prices* using the same notation like in [10], [6], [45] etc.

5.3.1 The Dual Problem

The first step in our solution is to define the Lagrangian function $L(x, p)$ for the optimization problem **P** as follows:

$$
\begin{aligned}
L(x, p) &= \sum_{m \in M} U_m(x_m) + \sum_{q \in Q} p_q(c_q - x_m \, \Gamma_{qm}) \\
&= \sum_{m \in M} \left(U_m(x_m) - x_m \sum_{q \in Q} p_q \Gamma_{qm} \right) + \sum_{q \in Q} p_q c_q
\end{aligned}
\tag{5.5}
$$

where $p = (p_q, q \in Q)$ is the vector of Lagrange multipliers. Notice that the first term in Equation (5.5) is separable in x_m, which means that

$$
\max_{x_m \in I_m} \sum_{m \in M} \left(U_m(x_m) - x_m \sum_{q \in Q} p_q \Gamma_{qm} \right) = \sum_{m \in M} \max_{x_m \in I_m} \left(U_m(x_m) - x_m \sum_{q \in Q} p_q \Gamma_{qm} \right)
$$

which means that this objective function can be divided into M subproblems. This decomposition of the problem solution into a set of subproblems each of which is easy to solve locally is a crucial advantage since it simplifies the overall solution and lays the base for a layered protocol design. The objective function of the dual problem then becomes (see Section 3.4.2 in [74])

$$D(p) = \max_{x_m \in I_m} L(x,p) = \sum_{q \in Q} p_q c_q +$$
$$\sum_{m \in M} \max_{x_m \in I_m} \left(U_m(x_m) - x_m \sum_{q \in Q} p_q \Gamma_{qm} \right) \tag{5.6}$$

and the dual problem **D** for the primal problem **P** can then be defined as follows:

$$\mathbf{D}: \quad \min_{p \geq 0} D(p) \tag{5.7}$$

It is important at this point to mention two properties of this dual problem from our model standpoint. First, since the objective function of problem **P** is concave, the constraints in Equation (5.3) are linear, and based on assumption 1 that there is at least one feasible solution for x, then there is no duality gap (Proposition 5.2.1 of [73]). Consequently, since there is no duality gap, the set of dual optimal prices (Lagrange multipliers) will always exist and will be equal to the set of Lagrange multipliers $p_q \forall q \in Q$ (Proposition 5.1.4 of [73]). This suggests that by solving the dual problem in Equation (5.7) we can get optimal dual variables (which are the Lagrange multipliers) $p_q^* \forall q \in Q$. Then, we will use this to solve the individual subproblems each with objective

$$\phi_m(x_m) = U_m(x_m) - x_m \sum_{q: (F_m \cap V_c^q) \neq \emptyset} p_q^* \Gamma_{qm} \quad \forall m \in M \tag{5.8}$$

Because these subproblems are separated, they can be solved for each multicast tree (e.g. by a designated tree member like the source of the tree) without coordinating with other multicast trees.

5.3.2 Interpretation of prices

Here, p_q may be interpreted as the *price* per unit bandwidth consumed at clique q. We can also define the aggregated price for group m as a result of consuming bandwidth on all maximal cliques $q \in Q$ as follows:

$$\lambda_m = \sum_{q:(F_m \cap V_c^q) \neq \emptyset} p_q \Gamma_{qm} \tag{5.9}$$

For each clique, the price is the product of the price per unit bandwidth of that clique and the number of *multicast subflows* of group m that exist in that clique. To calculate the price in Equation (5.9) for any group m in a centralized fashion we require one node to have knowledge about all cliques and the number of multicast subflows end-to-end in m that exist in each clique. Because this is not practical, we must devise a mechanism for calculating this group price in such a way that allows each node to contribute to the end-to-end calculation. The first step in this mechanism is to decompose the group price into individual subflow prices. In other words, the group price for group m will be the aggregated price of all the multicast subflows that belong to group m. The price of one multicast subflow is the summation of the clique prices that this multicast subflow belongs to. This indicates that this price is correlated with both the congestion on this subflow and its span which is measured by how many cliques this subflow exists in. The multicast subflow f_{mi} belongs to any clique q if at least one of its branches ($b_{mij} \; \forall j \in J$) belongs to that clique. In other words, two subflows belong to the same clique (contend) if at least one branch on each subflow has a source or a destination in the interference range of the other branch. The following example illustrates this part of the model.

Example for calculating group price

Consider the example in Figure 5.3. The network topology shown in Figure 5.3(a) has one multicast group m_1 which has two receivers (Nodes 5 and 6), and traffic that goes along four multicast subflows as shown in the figure. Figure 5.3(a) shows the contention graph corresponding to this network topology and the two maximal cliques with Prices p_1 and p_2 as shown. Based on Equation (5.9), the price for the multicast group m_1 is $\lambda_m = 3p_1 + 3p_2$, which is the sum of the product of the number of subflows of m_1 in each clique and the price of this clique. If we define the price for a subflow f_{mi} or λ_{mi} to be the total price of all the maximal cliques such that there is at least one branch of the subflow in that clique, i.e. $\lambda_{mi} = \sum_{q \in (f_{mi} \cap Q)} p_q$, then we can rewrite the group price as follows:

$$\lambda_m = \sum_{i:f_{mi} \in F_m} \lambda_{mi} = (p_1 + p_2) + (p_1 + p_2) + p_1 + p_2 \tag{5.10}$$

It is easy to see that Equation (5.10) is equivalent to (5.9). The reason for transforming the group price into this form is to facilitate the group price calculation in a distributed environment where we have all the nodes that are members of the group contributing in an accumulative fashion to the price calculation. Furthermore, for multicast group price calculation, we must consider all the branches of the multicast tree in the calculation because different subflows may exist in different branches of the tree. In the following sections, we will describe our approach for group price calculation which is a crucial part of our solution for resource allocation for wireless multicast.

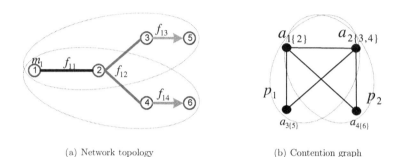

<table>
<tr><td>(a) Network topology</td><td>(b) Contention graph</td></tr>
</table>

Figure 5.3: Example for calculating the multicast group price.

5.4 Optimal resource allocation for wireless multicast (ORAWM)

First, we present a distributed iterative algorithm that solves the primal problem **P** by applying the gradient projection method [73] to the dual problem D. This implies that the clique prices $p_q^{(n+1)}$ at iteration $n + 1$ is adjusted in the opposite direction to the gradient $\nabla D(p)$ as follows:

$$p_q^{(n+1)} = [p_q^{(n)} - \alpha \frac{\partial D(p^{(n)})}{\partial p_q}]^+ \tag{5.11}$$

where $\alpha > 0$ is the step size, and $[z]^+ = \max\{z, 0\}$. Since U_m $m \in M$ are concave functions, $D(p)$ is continuously differentiable (pp. 669 in [74]) and hence its gradient based on Equation (5.6) is

$$\frac{\partial D(p)}{\partial p_q} = c_q - \sum_{m:(F_m \cap V_c^q) \neq \emptyset} x_m(p) \Gamma_{qm} \qquad q \in Q \tag{5.12}$$

which is the difference between the clique's achieved capacity and the load demanded on that clique. The clique load is characterized by the rates of all groups that belongs to this clique multiplied by the number of multicast subflows they have in this clique. Substituting Equation (5.12) into (5.11) we have

$$p_q^{(n+1)} = [p_q^{(n)} - \alpha(c_q - \sum_{m:(F_m \cap V_c^q) \neq \emptyset} x_m(p^{(n)}) \Gamma_{qm})]^+ \tag{5.13}$$

Although the dual objective in Equation (5.8) is not separable in p, the iterative algorithm in Equation (5.13) can be implemented in a decentralized way because only the local rates within each clique needs to be coordinated.

Finally, to calculate the rate for each group we can rewrite Equation (5.8) as a function of the group price as follows:

$$\phi_m(x_m) = U_m(x_m) - x_m \lambda_m \tag{5.14}$$

which represents the net benefit or the difference between the utility and the price for group m by allocating rate x_m. By Assumption 2, $\phi_m(x_m)$ is strictly concave and twice continuously differentiable, and hence a unique maximizer exits and can be obtained as follows:

$$\frac{d\phi_m(x_m)}{dx_m} = U'_m(x_m) - \lambda_m = 0$$

and hence the maximizer $x_m(\lambda_m^{(n)})$ at iteration t is

$$x_m(\lambda_m^{(n)}) = \begin{cases} w_m & \text{if } U_m'^{-1}(\lambda_m^{(n)}) \leq w_m \\ W_m & \text{if } U_m'^{-1}(\lambda_m^{(n)}) \geq W_m \\ U_m'^{-1}(\lambda_m^{(n)}) & o.w \end{cases} \qquad (5.15)$$

where w_m, and W_m are defined by the rate interval I_m.

5.4.1 Group price calculation

A crucial part of our algorithm is how to calculate the individual group prices λ_m in a decentralized way given the prices of the individual maximal cliques $p_q \quad \forall q \in Q$. To facilitate discussion in this part, we introduce the following new terms:

1. $\pi_m(i)$: the parent node of node i along the tree of group m.

2. $\lambda_m(i)$: the accumulated price for group m at node i..

Note that there is no parent node for the group source, i.e. $\pi_m(s_m) = \emptyset$, and the accumulated price at the source $\lambda_m(s_m) = 0$. We can then define the accumulated group price recursively as follows:

$$\lambda_m(i) = \frac{\lambda_m(\pi_m(i)) + \lambda_{m\pi_m(i)}}{K_{m\pi_m(i)}} \quad \forall i \in V_m \qquad (5.16)$$

where $K_{m\pi_m(i)}$ is the cardinality of subflow $f_{m\pi_m(i)}$. The reason for introducing the concept of accumulated group price will become evident when we discuss the different implementations of our algorithm (see Section 5.5.2). Equation (5.16) states that the accumulated price for group m at node i is the summation of the accumulated price at the parent node of i and the price for the multicast subflow $f_{m\pi_m(i)}$, divided then by the cardinality of this subflow $K_{m\pi_m(i)}$. This recursive algorithm will be the base for calculating the group price as illustrated by the following theorem.

Theorem 5.1. : *If $\Im_m \subseteq R_m$ such that $\Im_m = \{i : f_{mi} = \emptyset \; \forall i\}$ defines the set of terminal nodes for group m, then the group price can be calculated as follows:*

$$\lambda_m = \sum_{i \in \Im_m} \lambda_m(i) \tag{5.17}$$

Proof. Assume that $\Im_m(h)$ is the set of all nodes $i \in V_m$ of group m such that the depth of i is h, and H is the maximum depth of the tree of group m. Now it is easy to recognize that:

$$\lambda_m = \lambda_{ms_m} + \sum_{i \in \Im_m(1)} \lambda_{mi} + \cdots + \sum_{i \in \Im_m(H-1)} \lambda_{mi}$$

We proceed by induction based on H as follows:

- For $H = 1$:

$$\sum_{i \in \Im_m} \lambda_m(i) = \sum_{i \in \Im_m(1)} \lambda_m(i) = K_{ms_m}.\lambda_{ms_m}/K_{ms_m} = \lambda_m$$

- For $H = 2$:

$$\sum_{i \in \Im_m(2)} \lambda_m(i) = \sum_{i \in \Im_m(2)} \frac{\lambda_{ms_m}/K_{ms_m} + \lambda_{m\pi_m(i)}}{K_{m\pi_m(i)}} = \sum_{i \in \Im_m(1)} \frac{\lambda_{ms_m}/K_{ms_m} + \lambda_{mi}}{K_{mi}}.K_{mi}$$

$$= \lambda_{ms_m} + \sum_{i \in \Im_m(1)} \lambda_{mi} = \lambda_m$$

Notice that if $f_{mi} = \emptyset$, then $\lambda_{mi} = 0$.

- Assume that for $H = n - 1$:

$$\sum_{i \in \Im_m(n-1)} \lambda_m(i) = \lambda_m = \lambda_{ms_m} + \cdots + \sum_{i \in \Im_m(n-2)} \lambda_{mi}$$

- Hence, for $H = n$:

$$\sum_{i \in \Im_m(n)} \lambda_m(i) = \sum_{i \in \Im_m(n-1)} \frac{\lambda_m(i) + \lambda_{mi}}{K_{mi}} . K_{mi} = \lambda_{ms_m} + \cdots + \sum_{i \in \Im_m(n-1)} \lambda_{mi} = \lambda_m$$

and therefore the result follows. $\square$

Theorem 5.1 implies that in order to calculate the price for group m, we need to calculate the accumulated price on each branch using Equation (5.16) until we hit the terminal node of this branch. Then, send this accumulated price back to the source (or any other designated group member) to calculate the group price by simply adding the accumulated price of all branches. This simple mechanism for calculating the group price facilitates our solution due to the following features:

1. It is simple to implement using one header field that stores the accumulated price for all branches of subflow f_{mi} (i.e. note from Equation (5.16) that the accumulated price is the same for all branches that belong to the same subflow).

2. It allows for distributed computation because each node needs to know (for each group) only the accumulated price for its parent node and the price of the next subflow. Then, it performs a simple calculation to get the accumulated price for the following nodes.

3. It also allows for asynchronism because the source does not need to wait for all accumulated prices from all branches to calculate the group price and the next

group's optimal rate. Instead, it uses estimation based on the previously received values as will be shown next.

Now we illustrate the result of Theorem 5.1 by an example. Consider the network topology in Figure 5.3. We can derive the accumulated price at Nodes 5, and 6 based on Equation (5.16) as follows:

$$\lambda_1(5) = (\lambda_{11} + \lambda_{12})/2 + \lambda_{13} = 2p_1 + p_2$$

$$\lambda_1(6) = (\lambda_{11} + \lambda_{12})/2 + \lambda_{14} = 2p_2 + p_1$$

and we can easily see that $\lambda_m = \lambda_1(5) + \lambda_1(6) = 3p_1 + 3p_2$.

5.4.2   ORAWM: Synchronous Distributed algorithm (ORAWM_SD)

Now we describe the first version of our distributed algorithm where we assume that all the calculations happen in synchronization at specific time instances. Figure 5.4 depicts the main procedures performed in the network for calculating the optimal rates for each multicast group.

This algorithm works by assuming that for each clique in the network, there is a representative node that has the topology information for the entire clique. This node will receive the rates for all the active multicast groups and calculate the clique price based on that. This node will then communicate this price to all the nodes within that clique to perform the subflow price calculations. In order to prove the convergence of

Clique procedure (by clique q): At iterations $n = 1, 2, \cdots$

1. Receive rates $x_m^{(n)}$ from all groups m where $F_m \cap V_c^q \neq \emptyset$

2. Update clique price as follows

$$p_q^{(n+1)} = [p_q^{(n)} - \alpha(c_q - \textstyle\sum_{m:(F_m \cap V_c^q) \neq \emptyset} x_m(p^{(n)})\Gamma_{qm})]^+$$

3. Send $p_q^{(n+1)}$ to all nodes of group m such that $F_m \cap V_c^q \neq \emptyset$

Subflow procedure (by subflow f_{mi}): At iterations $n = 1, 2, \cdots$

1. Receive prices from all maximal cliques q where $f_{mi} \cap V_c^q \neq \emptyset$

2. Calculate the subflow price (per hop price) λ_{mi} as follows

$$\lambda_{mi} = \textstyle\sum_{q:(f_{mi} \cap V_c^q) \neq \emptyset} p_q^{(n)}$$

3. Calculate the accumulated price on each branch $b_{mij} \in f_{mi}$

$$\lambda_m(j) = \frac{\lambda_m(i) + \lambda_{mi}}{K_{mi}} \quad \forall j \in J$$

4. Forward $\lambda_m(j)$ to all children subflows of f_{mi}, if no children, send back $\lambda_m(j)$ to s_m

Group procedure (by group source s_m): At iterations $n = 1, 2, \cdots$

1. Receive accumulated prices $\lambda_m(i)$ from all branches of m $\forall i \in \Im_m$

2. Calculate the group price λ_m as follows

$$\lambda_m = \textstyle\sum_{i \in \Im_m} \lambda_m(i)$$

3. Calculate the next group rate as follows

$$x_m^{(n+1)} = U_m'^{-1}(\lambda_m^{(n)})$$

4. Send $x_m^{(n+1)}$ to all cliques q where $F_m \cap V_c^q \neq \emptyset$

Figure 5.4: ORAWM: Synchronous Distributed algorithm.

this algorithm we define the following new terms:

1. $Y_m = \sum_{q \in Q} \Gamma_q m$ which measures the number of subflows in clique q for group m, and

   $\bar{Y} = \max_{m \in M} \sum_{q \in Q} \Gamma_q m$ indicates the number of subflows for the biggest group $m \in M$.

2. $Z_q = \sum_{m \in M} \Gamma_q m$ which measures the number of subflows of group m in clique q, and

   $\bar{Z} = \max_{q \in Q} \sum_{m \in M} \Gamma_q m$ indicates the number of subflows in the most congested clique

   $q \in Q$.

3. $\bar{\gamma} = \max_{m \in M} \gamma_m$ indicates the upper bound on all $-U''_m(x_m) \ \forall m \in M$.

Now, assuming that the initial price for each clique is feasible, i.e. $p_q \geq 0 \forall q \in Q$, we can

obtain the following convergence result:

Theorem 5.2. : *For step size values of α that satisfy the inequality $0 < \alpha < 2\bar{\gamma}/\bar{Y}\bar{Z}$,*

starting from any initial rate $x(0)$ ($x_m \in I_m \ \forall m \in M$), and clique prices $p(0) \geq 0 \ \forall q \in Q$,

every accumulation point $(x^*; p^*)$ of the sequence $(x(n); p(n))$ generated by the Algorithm

in Figure 5.4 is primal-dual optimal.

The proof is shown in details in Appendix B.

5.4.3   ORAWM: Asynchronous Distributed algorithm

(ORAWM_AD)

Our **ORAWM_SD** algorithm in Figure 5.4 assumes that updates at the cliques, subflows,

and groups are synchronized which means that each of the iterations $n = 1, 2, \cdots$ occurs at the same time in all three procedures. This restricts the calculations to be performed by minimum number of nodes in order to guarantee the synchronization amongst these nodes. Thus, it is better to elect a node within each clique to act as the master node that performs clique price calculations. Similarly, elect a group representative node (group source, or group leader as we will see later) that performs the group price based on which the next group rate is updated. Such synchronization, however, is difficult to achieve even between few number of nodes and requires lots of packet overhead to convey the synchronization information. Besides, the clique master election process is also difficult to achieve in real ad hoc networks because of mobility. As network nodes start to move, the clique topology changes and hence the master node of the clique may change. This calls for a modification to this algorithm that works asynchronously on all the group members in such a way that distributes the calculations without any synchronization requirements.

In the following, we present the asynchronous version of our algorithm (**ORAWM_AD**) and prove its convergence. This asynchronous model and the proof of convergence follow the class of partially asynchronous algorithms which were first discussed in Chapter 7 of [74]. For this partially asynchronous algorithm, we still assume that each clique has a master node denoted as v_q, which performs the *clique procedure* through which a node calculates the estimated clique price and sends it to all nodes which have a subflow belonging to that clique. In our implementations, however, we will show that even if the clique price calculations are performed by each node independently and asynchronously, convergence is still attainable. Each node which is a member of any multicast group $m \in M$ performs the *subflow procedure* by receiving the clique prices for all cliques such that node i has a subflow f_{mi} which belongs to that clique. This node then calculates the accumulated price through the recursive Equation (5.16), and forwards this price to children nodes. Finally, as we hit a terminal node of group $m \in \Im_m$, the accumulated price for that branch is fed back to the group source for preforming the *group procedure* through which the group source calculates the group price, and hence the next group rate. We also make the following assumption for the maximum delay bound between any two nodes:

Assumption 3: For all groups $m \in M$, cliques $q \in Q$, and nodes $i \in M$, the time between consecutive updates is at most B time units.

Now, let $T = \{0, 1, 2, \cdots\}$ be the set of times at which either the rates or the prices are updated at any node. Hence, we define some time instance terminology as follows:

1. $T_q \subseteq T$: the set of time instances at which master node $v(q)$ updates p_q.

2. $T_m^i \subseteq T$: the set of time instances at which node $i \in M$, which has $f_{mi} \in F_m$ updates λ_{mi} and $\lambda_m(i)$.

3. $T_m \subseteq T$: the set of time instances at which the source of group m updates x_m.

The main idea of this asynchronous algorithm, is that for any update at any node, the node may not have the exact current value of either the rate or the price. Instead, it receives a sequence of recent values at different time instances. Therefore, the node will use a weighted average of these values in estimating the price or the rate at any given time. The details for our asynchronous algorithm (**ORAWM_AD**) are shown in Figure 5.5.

This asynchronous model is general and can allow for any update policy for the rates or prices. For example, the accumulated price fed back to the source from some terminal nodes may be delayed or lost for some $\lambda(i, \tau)$ $\tau \in [t' - B, \ldots, t]$, but we can still guarantee the convergence by estimating the accumulated price based on the recent received values. Also, the algorithm attains convergence using some popular update policies such as the following:

- **Latest instant update:** only the last received clique price $p_q(\tau)$ for some $\tau \in [t - B, \cdots, t]$ is used to estimate $\hat{\lambda}_m(t)$, i.e. $b_i^q(t', t) = b_q^i(t', t) = b_m^i(t', t) = 1$ if $t' = \tau$ and 0 otherwise.

- **Latest average update:** only the average over the latest k received values is used for estimation, i.e. $b_i^q(t', t), b_q^i(t', t), b_m^i(t', t) > 0$ for $t' = \tau - k + 1, \cdots, \tau$ and 0 otherwise.

The support for these different update policies demonstrates the versatility of our asynchronous algorithm. The following theorem states this convergence result.

Theorem 5.3. : *For step size values of α that are sufficiently small, starting from any initial rate $x(0)$ ($x_m \in I_m$ $\forall m \in M$), and clique prices $p(0) \geq 0$ $\forall q \in Q$, every accumulation point $(x^*; p^*)$ of the sequence $(x(t); p(t))$ generated by the asynchronous*

Clique procedure (by clique q): At times $t \in T_q$

1. Receive rates $x_m(t')$ from all groups m where $F_m \cap V_c^q \neq \emptyset$ and keep $x_m(t')$ for $(t-B) \leq t' \leq t$

2. Estimate the rate $\hat{x}_m(t)$ as follows:

$$\hat{x}_m(t) = \sum_{t'=t-B}^{t} b_i^q(t', t) \, x_m(t') \quad \text{with} \quad \sum_{t'=t-B}^{t} b_i^q(t', t) = 1$$

3. Update clique price as follows
$$p_q(t+1) = [p_q(t) - \alpha(c_q - \sum_{m:(F_m \cap V_c^q) \neq \emptyset} \hat{x}_m(t)\Gamma_{qm})]^+, \ \forall t \in T_q \text{ and } p_q(t+1) = p_q(t) \ \forall t \notin T_q$$

4. Send $p_q(t+1)$ to all nodes of group m such that $F_m \cap V_c^q \neq \emptyset$

Subflow procedure (by subflow f_{mi}): At times $t \in T_m^i$

1. Receive prices $p_q(t')$ from all maximal cliques q where $f_{mi} \cap V_c^q \neq \emptyset$
   and keep $p_q(t')$ for $(t-B) \leq t' \leq t$.

2. Estimate the clique price $\hat{p}_q(t)$ as follows:

$$\hat{p}_q^i(t) = \sum_{t'=t-B}^{t} b_q^i(t', t) \, p_q(t') \quad \text{with} \quad \sum_{t'=t-B}^{t} b_q^i(t', t) = 1$$

3. Calculate the subflow price (per hop price) λ_{mi} as follows

$$\lambda_{mi}(t+1) = \sum_{q:(f_{mi} \cap V_c^q) \neq \emptyset} \hat{p}_q(t)$$

4. Calculate the accumulated price on each branch $b_{mij} \in f_{mi}$
   $$\lambda_m(j, t+1) = \frac{\lambda_m(i,t) + \lambda_{mi}(t+1)}{K_{mi}} \quad \forall j \in J \quad \forall t \in T_m^i \text{ and } \lambda_m(j, t+1) = \lambda_m(j,t) \ \forall t \notin T_m^i$$

5. Forward $\lambda_m(j, t+1)$ to all children subflows of f_{mi}, if no children, send back $\lambda_m(j, t+1)$ to s_m

Group procedure (by group source s_m): At times $t \in T_m$

1. Receive accumulated prices $\lambda_m(i, t')$ from all branches of m $\forall i \in \Im_m$,
   and keep $\lambda_m(i, t')$ $\forall i \in \Im_m$ and $(t-B) \leq t' \leq t$.

2. Estimate the accumulated price for each terminal node $i \in \Im_m$ as follows

$$\hat{\lambda}_m(i, t) = \sum_{t'=t-B}^{t} b_m^i(t', t) \lambda_m(i, t') \quad \text{with} \quad \sum_{t'=t-B}^{t} b_m^i(t', t) = 1$$

3. Calculate the next group price $\lambda_m(t+1)$ as follows

$$\lambda_m(t+1) = \sum_{i \in \Im_m} \hat{\lambda}_m(i, t)$$

4. Calculate the next group rate as follows
   $$x_m(t+1) = U_m'^{-1}(\lambda_m(t+1)) \ \forall t \in T_m \text{ and } x_m(t+1) = x_m(t) \ \forall t \notin T_m$$

5. Send $x_m(t+1)$ to all cliques q where $F_m \cap V_c^q \neq \emptyset$

Figure 5.5: ORAWM: Asynchronous distributed algorithm.

algorithm in Figure 5.5 is primal-dual optimal.

The proof of this theorem is in Appendix B.

5.4.4 Time-varying environment

The two algorithms discussed so far, namely: **ORAWM_SD**, and **ORAWM_AD** with their convergence analysis, assume that the cliques' achieved capacity and the set of group utility functions are not functions of time (i.e. they do not change with time). However, due to online calculation, and subproblem decomposition, it can be shown that the algorithms work in the case when these quantities change with time.

For example, the clique's capacity may be time varying depending on the scheduling discipline used at the MAC layer. In this case, the *clique procedure* in Figure 5.5 will be the same except that in computing $\nabla D(p(t))$ at step 3, the current clique capacity $c_p(t)$ is used in place of the constant capacity c_p. The same applies to the time-varying utility functions $U_m(x_m, t)$. In general, if the change in the environment parameters is slow relative to the convergence rate of the algorithm, the algorithm can track the changes in the optimal rates based on changing these quantities with time. This is shown in our experimental evaluation discussed in the following section.

5.5 Experimental Evaluation

In this section, we present a group of implementations for our algorithms in different wireless networking environment settings. We evaluate the performance of our algorithms using a set of simulation environments implemented using the ns-2 simulator.

5.5.1 Implementations overview

We study the performance of our wireless multicast algorithm in different simulation environments as depicted in Figure 5.6. For convenience, we divide the implementations into two categories. The *information distribution* category defines three main simulation environments based on how price and rate information are handled by the network nodes. The first environment, referred to as **ORAWM_CS**, assumes that all the price and rate calculations for all multicast groups are performed by a designated node. This means that this node must have the topology information to construct all the maximal cliques and calculate the group prices for all multicast groups. Although such an environment is not practical and it poses many deployment issues especially in large networks, it is a straightforward way of guaranteeing synchronization, and hence would be interesting to compare the asynchronous environments against it. The second environment, referred to as **ORAWM_PDA**, implements the algorithm in Figure 5.5, assuming a clique master node for each clique, which performs the clique price calculations asynchronously as described before. Although this environment is more practical than the first one because of its asynchronous nature, it may still pose some deployment difficulties because of its

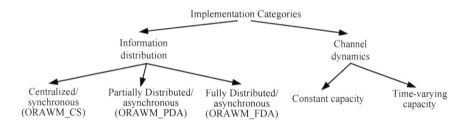

Figure 5.6: Simulation environments.

intra-clique centralized calculations. Communication between the master node and all other nodes within a clique each time the price changes causes some communication overhead. Also, electing a master node for each clique each time the topology changes may be complicated especially for networks with mobility or frequently changing topology. In the third environment, referred to as **ORAWM_FDA**, each node performs the clique price calculation independently based on the topology information it has. In [6], it is shown that, for clique construction, the status of all wireless links should be reported to all nodes as far as *three* hops away. This will enable all nodes to completely determine all the cliques containing any of its adjacent links [3] and hence the node will be able to perform clique and subflow price calculations without having to coordinate prices with other nodes. This eliminates the communication overhead required to convey clique price information and does not require any intra-clique centralized calculations.

Each of these three simulation environments can work under one of two network conditions illustrated by the *channel dynamics* category in Figure 5.6. One scenario

---

[3]Notice that this result applies directly to the clique construction for the multicast subflows.

Table 5.1: MAC and physical layer parameters used by our simulations.

Transmission range		250m
Interference range		550m
Radio propagation model		Two-ray ground reflection
Channel	basic rate	1Mbps
parameters	data rate	1Mbps
Simulation wireless area		$1000 \times 1000m^2$

assumes that the MAC layer scheduling is *ideal* in the sense that it can always achieve the wireless channel capacity (i.e. channel utilization is 1). The other scenario is more practical, because it can track the changes in the channel capacity regardless of the scheduling technique used in the MAC layer using window measurement techniques such as the one presented in [75]. Hence, it can work with almost any MAC layer scheduling. For this scenario, we use the common IEEE 802.11 DCF as the MAC protocol with multicast extensions as presented in [71]. Table 5.1 summarizes the MAC and physical layer parameters used by our simulation environments.

5.5.2 Solution architecture and deployment

Deploying our solution to real networks requires cross layer design and entails some modifications to the MAC, network, and transport layers. Figure 5.7 depicts the cross-layer architecture of our solution **ORAWM** based on the algorithm in Figure 5.5. The figure shows the main procedures of **ORAWM** (on the right hand side) and its interaction with the different layers including MAC, routing and transport layers. In the following, we highlight the modifications and show the interaction of **ORAWM** in each layer.

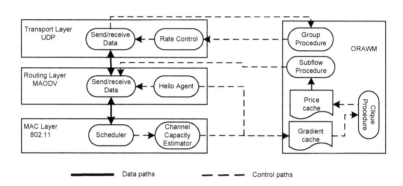

Figure 5.7: Cross-layer architecture for **ORAWM**

1. MAC layer: Unlike unicast, existing IEEE 802.11 MAC protocol does not provide any media access control recovery on multicast/broadcast packets. This means that the RTS/CTS/ACK mechanism is not used for multicast/broadcast packets from one sender to all the receivers on the same multicast subflow. As a result, the reliability of the multicast/broadcast service is reduced due to the increased probability of lost packets resulting from interference or collisions. For time-varying channel capacity at different contention regions (cliques), a reliable deterministic technique is required for channel capacity estimation. The channel capacity estimation for multicast packets poses a real challenge because it requires monitoring the multicast packets on both the sender and all the receivers of the subflow, and hence it calls for a synchronized clock to measure the delay of sending the multicast packet from the sender to all the receivers. Because this is not a practical solution, we devise a deployment technique which combines the multicast aware MAC protocol (MMP) [71] with a bandwidth management mechanism

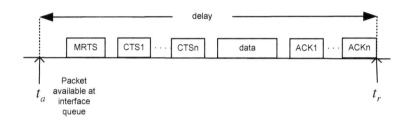

Figure 5.8: Multicast-aware MAC protocol.

for measuring the channel capacity based on [75].

First, MMP provides a MAC layer support for multicast traffic by attaching an Extended Multicast Header (EMH) which combines the address of the nexthop nodes using some routing information. The MAC layer then uses the EMH field to support an ACK based data delivery from the sender to all the receivers on the same multicast subflow. After sending the data packet, the transmitter waits for the ACK from each of its destinations in a strictly sequential order (hence, avoids the contention between the ACK packets on the sender side). A retransmission of the multicast packet is performed only if the ACK from any of the nodes in EMH is missing. The retransmission is done using a similar technique like RTS/CTS, but this time using Multicast RTS or MRTS/CTS. Figure 5.8 shows the full scenario from the sender side for an MRTS/CTS/data/ACK from one sender to n receivers.

Second, bandwidth estimation can then be performed using a measurement-based window technique such as the one presented in [75] and adopted by [6]. In this technique, the achievable bandwidth of each *multicast subflow* is measured based on its recent received packets at the sender of this subflow. As shown in Figure 5.8, the packet becomes *available* at the MAC layer interface queue at time t_a, and when all the acknowledgments are received at the sender node, we claim that the packet is received at time t_r. The transmission delay over this subflow is then $t_r - t_a$, which includes the contention period, and retransmissions (if any). We then use a window w of packets to measure the achievable bandwidth observed by the subflow as follows $B(f_{mi})$:

$$B(f_{mi}) = \frac{w.z}{\sum_{i=1}^{w} t_r^i - t_a^i}$$

where z is the packet size.

2. Network layer: The price and rate information are exchanged in the same level as the routing information. We have integrated our solution with the Multicast Ad hoc On demand Distance Vector (MAODV) presented in [76] for providing a distributed routing scheme for the multicast sessions. MAODV extends (AODV) [77] to offer multicast capabilities by building multicast trees as needed and providing tree operations (e.g. merge, and prune trees) for tree maintenance in a distributed fashion. For clique construction and price calculation, we use a multicast-based status which is conveyed in a *gradient information header* (gih) for each aggregated subflow within the clique. To accommodate network changes (i.e. changes due to mobility or source start/stop), this multicast-based status is stored using a caching mechanism in each node with expiry timestamp that controls its lifetime. We piggyback this status information onto the HELLO packets and allow it to be transmitted as far as three hops away to enable each node to construct the clique and calculate the clique price. For clique construction, we use the Bierstone algorithm described in [78]. The accumulated price and group rate are also piggybacked within the *price information header* (pih) onto the data packets. The tree terminal nodes notify the group source of the new accumulated price by sending a *feedback packet* (fbp) as shown in Figure 5.9. This feedback packet is then used by the source to calculate the group price as explained in Section 5.4.3. These feedback packets need not be synchronized and in fact even if some of them are delayed or lost, the source can still estimate the group price using the estimation procedure highlighted in Figure 5.5.

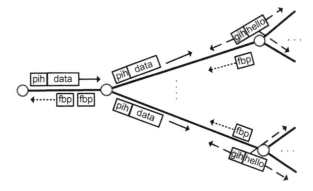

Figure 5.9: Price and rate control packets.

3. Transport layer: Our algorithm provides an end-to-end flow control in addition to optimizing network resources because it performs online control to the rates used by each transport agent of the multicast groups. Therefore, to minimize the communication overhead we use UDP as the transport protocol and we add the rate and price adjustment as part of sending and receiving data. In particular, we add the functionality of sending a feedback packet asynchronously from the terminal nodes every time the accumulated price changes (i.e. before convergence happens). We also modify the sender transport agent to adjust the sending rate based on the summation of the accumulated prices from all terminal nodes [4] . Using the result from Theorem 5.3, the sender can calculate the rate asynchronously based on the estimated value of the accumulated prices from all terminal nodes every time it receives a feedback packet from one of these terminal nodes.

5.5.3 Simulation results

In all our experiments, we use the utility functions $U_m(x_m) = g_m \ln(x_m)$ $x_m > 0$ for imposing proportional fairness amongst the multicast groups where g_m is the differentiation gain, i.e. $x_m(t) = g_m/\lambda_m(t)$. Unless otherwise stated, we use the latest instant update for asynchronous calculations.

Convergence for different simulation environments

We first study the convergence of our **ORAWM** algorithm in the different implementations explained in Section 5.5.1. We take as an example the network in Figure 5.2 with 3 multicast sessions as shown in Figure 5.2(c) with equal differentiation gain, i.e. $g_m = 1$ $\forall m \in M$. However, we start each of these sessions in a different time to test the ability of our algorithm to track network changes. The start times of sessions m_1, m_2, and m_3 are $20, 40$, and 60 seconds respectively, and the initial rates $x_m(0)$ $\forall m \in M$ are

---

[4]The termination criteria in all our simulations are $|\lambda_m(t) - \lambda_m^*| \leq \mu$ $\forall m \in M$ with $\mu = 10^{-5}$

selected from a uniform distribution in the range [50, 250] kbps. We fixed all the other parameters including the step size, and we measured the rate of each multicast session against time. Figures 5.10 and 5.11 show the results for fixed channel and time-varying channel capacity respectively.

We notice in all these cases that our algorithm converges in the three environment settings and regardless of any initial settings, or the initial rates set in each case. We notice that the only difference between **ORAWM_CS** and **ORAWM_PDA** implementations is in the transient value of the rates against time. However, neither the convergence speed nor the optimal rates are affected by the asynchronism that we introduced by the **ORAWM_PDA** implementation, which confirms the result we stated in Theorem 5.3.

We notice that convergence also happens using the **ORAWM_FDA** implementation with occasionally very small error in the calculated optimal rates as shown in Figures 5.10(c) and 5.11(c). This small error happens because the clique price calculation is performed by each node within the clique independently based on the current status of the subflows. This small error, however, seems to be acceptable and should not increase by increasing the number of hops since it is localized within each clique. In Figure 5.11, we observe that although the MAC channel capacity (i.e. the basic rate of sending data in IEEE 802.11 DCF) is set to 1Mbps, the achieved channel capacity changes with time and does not go above 800Kbps. Nevertheless, our algorithm continuously tracks the change in channel capacity and provides proportional fairness amongst all the multicast sessions based on the current available channel capacity.

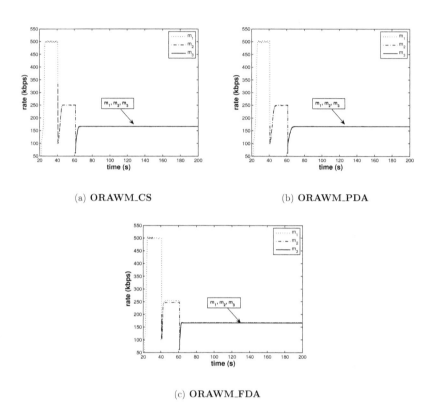

(a) **ORAWM_CS** (b) **ORAWM_PDA**

(c) **ORAWM_FDA**

Figure 5.10: Convergence for different simulation environments using fixed channel capacity.

Effect of step size α on convergence using the ORAWM_FDA implementation

In the previous experiment, we set α so that the convergence rate is in the best way possible. Now, we study the impact of this step size on the convergence using the

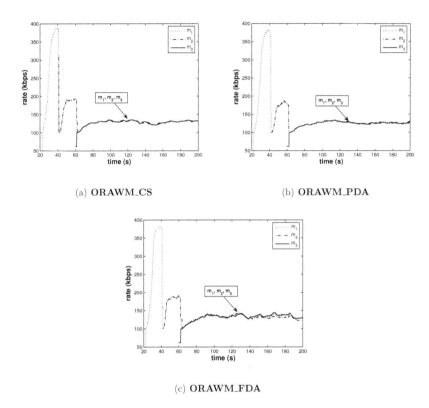

(a) **ORAWM_CS**

(b) **ORAWM_PDA**

(c) **ORAWM_FDA**

Figure 5.11: Convergence for different simulation environments using dynamic channel capacity.

ORAWM_FDA implementation. We take the same network explained in the previous experiments, we start the three sessions at time $t = 20$ seconds, and we set the differentiation factors a_1, a_2, and a_3 as 3, 2, and 1 respectively. Figure 5.12 shows the convergence behavior at different values of α. It is clear that our asynchronous algorithm

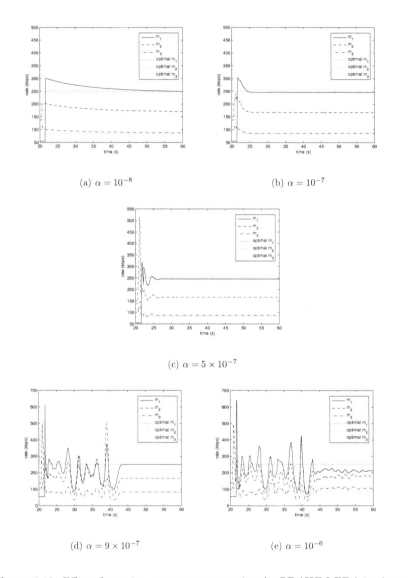

(a) $\alpha = 10^{-8}$

(b) $\alpha = 10^{-7}$

(c) $\alpha = 5 \times 10^{-7}$

(d) $\alpha = 9 \times 10^{-7}$

(e) $\alpha = 10^{-6}$

Figure 5.12: Effect of step size α on convergence using the **ORAWM_FDA** implementation.

achieves the global optimal rates for small values of α. As intuitively expected, however, for very small values of $\alpha < 10^{-7}$, the convergence rate tends to be slow (see Figure 5.12(a)). For large values of α (e.g. $\alpha = 10^{-6}$), the algorithm may not converge to the optimal rates.

Effect of estimation window B on convergence using the ORAWM_FDA implementation

So far, we have used the *latest instant update* for asynchronous calculations. This means that, we only consider the latest value when we estimate the clique and group prices. In this experiment, we study the effect of using an estimation window B seconds, and estimate the prices based on the average in this window (i.e. $b_i^q(t', t) = 1/n, b_q^i(t', t) = 1/n'$, and $b_m^i(t', t) = 1/n'' \; \forall i \; \forall q \; \forall m$, where n, n', and n'' are the number of received values within the window B). We notice from the results in Figure 5.13, that the convergence occurs in all values of B. However, by increasing B the convergence tends to happen with less fluctuations in the transient rates before they settle on the global optimal values. Hence, using a moderate estimation window might be useful, especially for large multicast trees, because some receivers may get overwhelmed by these rate fluctuations. We also notice that using large values of B (see Figure 5.13(c)) may affect the speed of convergence, and hence large estimation windows also might not be desirable.

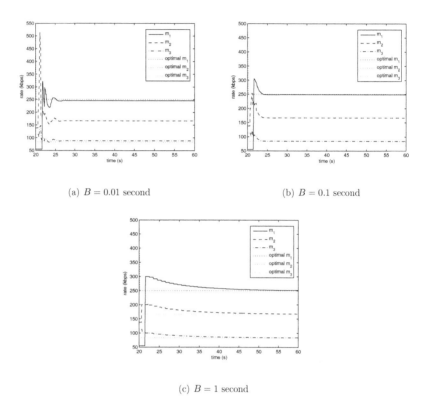

(a) $B = 0.01$ second   (b) $B = 0.1$ second

(c) $B = 1$ second

Figure 5.13: Effect of estimation window B on convergence using the **ORAWM_FDA** implementation.

Effect of differentiation gain and session start/stop time on convergence using the ORAWM_FDA implementation

In this experiment, we use fixed channel capacity, assuming that the scheduling mechanism used in the MAC layer is ideal (i.e. channel utilization is always one). In addition to the session start times we set the stop time for sessions m_1, m_2, m_3 to $120, 140, 160$ seconds respectively. We show the difference between the rates obtained using our algorithm running in a totally distributed and asynchronous environment, and the optimal theoretical rates calculated offline. Figure 5.14 shows the results using differentiation gains 3, 2, and 1 respectively. We notice that the algorithm converges online to the optimal rates that guarantees maximum aggregated utility providing proportional fairness amongst all active multicast sessions at any given time. Furthermore, it responds fairly quickly to the changes resulting from adding or removing multicast sessions.

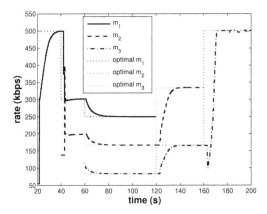

Figure 5.14: Convergence with different differentiation gains and start/stop times.

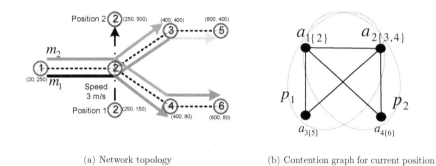

(a) Network topology   (b) Contention graph for current position

Figure 5.15: Impact of mobility and route changes.

Effect of mobility and route changes on convergence using the ORAWM_FDA implementation

In this experiment, we study the impact of mobility and route changes on our algorithm using the network in Figure 5.15(a). The corresponding contention graph for this case is shown in Figure 5.15(b). We generate a mobility pattern where Node 2 moves from *Position 1* to *Position 2* as shown in the figure using a speed of 3 m/s and a pause time of 20 seconds. We start 2 multicast sessions m_1 and m_2 with receiver sets [4, 5] and [3, 6] respectively at time 20 seconds. Figure 5.16 shows the rates calculated by our algorithm and the throughput observed at receivers $[r_4, r_5, r_3, r_6]$ respectively. The figure shows 3 different regions depending on the change of routes resulting from the node mobility. In Region 1, only node 4 is receiving traffic for group m_1 and node 6 is receiving traffic for group m_2. As expected in this case, our algorithm discriminates against the flow with

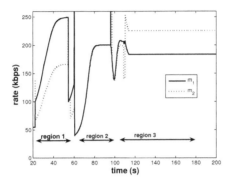

(a) Calculated group rate

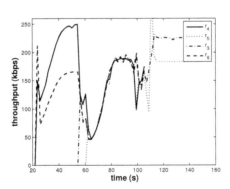

(b) Receiver throughput

Figure 5.16: Convergence with mobility and route changes.

longer path because it consumes more network resources, giving Group m_1 higher rate in this case (note that this discrimination can be compensated, if desired, by changing the utility functions). As Node 2 moves to Region 2, the routes for which are shown by Figure 5.15(a), all receivers become active and the optimal rates converge to the same value for both groups. The scenario in Region 1 will then be reversed in Region 3 giving m_2 higher rate in this case.

Convergence in random network using the ORAWM_FDA implementation

In this experiment, we study the convergence behavior of our algorithm **ORAWM** with respect to both calculated rate and throughput in a randomly generated wireless network as shown in Figure 5.17. This network consists of 30 nodes deployed randomly over the $1000 \times 1000 \ m^2$ wireless space. We started 3 multicast sessions m_1, m_2, and m_3 at time 20 seconds, each with one source s_i and four receivers as shown in Table 5.2 using $\alpha = 10^{-8}$. The utility functions for all the three sessions is $\ln(x_m)$ (i.e. $g_m = 1 \ \forall m \in M$).

Figure 5.18 shows the calculated rates and receiver throughput of each multicast session with time. From these results, we observe that our algorithm attains convergence with satisfactory speed even in relatively large scale networks. We also observe that the throughput achieved by each receiver on all sessions follows the calculated rates fairly well, which confirms the correctness of the calculated rates. Note that the optimal calculated rates are different for each session depending on the size of the multicast tree and how much resources each session consumes from the total network resources. If this discrimination based on tree topology is undesirable, it can be compensated using

different differentiation gains (g_m) on each session.

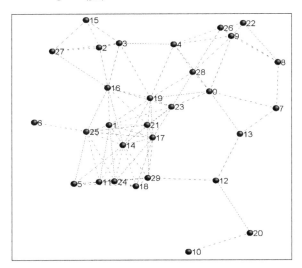

Figure 5.17: Random wireless network with 30 nodes.

Table 5.2: Multicast traffic pattern.

Session	Source	Receivers
m_1	s_0	$r_{11}, r_{12}, r_{13}, r_{14}$
m_2	s_2	$r_{16}, r_{17}, r_{18}, r_{19}$
m_3	s_4	$r_{26}, r_{27}, r_{28}, r_{29}$

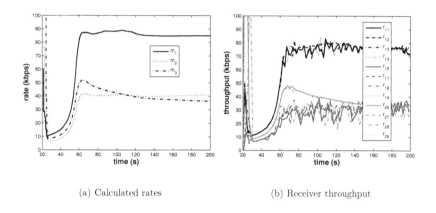

(a) Calculated rates (b) Receiver throughput

Figure 5.18: Impact of mobility and route changes on convergence of **ORAWM**.

Overhead study using the ORAWM_FDA implementation

In this experiment, we evaluate the overhead of our algorithm **ORAWM** using different network topologies. We measure the aggregate utility and normalized overhead for eight different topologies generated randomly, each with 30 nodes. We start three multicast sessions at time 20 seconds with a traffic pattern as shown in Table 5.2. The aggregate utility is defined in Equation (5.2) and the normalized overhead is the ratio between the number of non-data packets and data packets delivered at each hop. This overhead includes the FEEDBACK packets sent by the receivers of each multicast session (i.e. fbp packet), and the AODV routing packets, which include the HELLO packets that carry the price information (i.e. gih header) and the price information header (i.e. pih header). Note that the FEEDBACK packets are sent by the terminal node only if the accumulated price on that node changes (prior to convergence) in order to decrease the overhead. Furthermore, we compare the aggregate utility and overhead of **ORAWM** with two other heuristic algorithms. In these heuristic algorithms, the gih header is allowed to be transmitted either for one or two hops away instead of three as **ORAWM** does in order to reduce the overhead. In this case, the network nodes have only partial knowledge of the network topology and load, and hence, the constructed cliques and prices of them are computed approximately.

Figure 5.19 shows the aggregate utility and normalized overhead for the three algorithms. We notice in Figure 5.19(a) that the optimal algorithm (**ORAWM**) outperforms the other two algorithms consistently for all topologies as expected whereas the difference in aggregate utility between the other two algorithms is insignificant. An interesting result is shown in Figure 5.19(b) where the total overhead for the optimal algorithm is less than the 2-hop algorithm for some topologies (e.g. topology number 6). This is because more FEEDBACK packets are sent in the 2-hop case because of the approximate clique construction and price calculation, which increases the overall overhead.

5.6   Concluding Remarks

In this chapter we have presented a new multicast-based algorithm and analytical model for resource optimization over multihop ad hoc networks. This algorithm is used to control the rates of the multicast sessions in such a way that guarantees the optimal resource utilization of the wireless network resources while achieving fairness amongst the

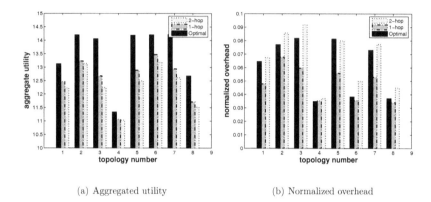

(a) Aggregated utility

(b) Normalized overhead

Figure 5.19: Aggregated utility and packet overhead for eight random networks each with 30 nodes.

multicast sessions utilizing the bandwidth-efficiency feature of multicast, which increases the overall network available bandwidth. Based on using maximal cliques as the main resource entity in the network, we modeled the multicast sessions in a form of contention domains and calculated the group price for the end-to-end multicast session. We proposed a mechanism for calculating the group price based on the branch accumulated price which allows the calculation to occur in a totally distributed and asynchronous way. We also demonstrated our theoretical claims and proofs by a series of simulations designed to capture different aspects of the wireless network conditions.

Chapter 6

Optimal Resource Allocation for Heterogeneous Wireless Multicast

In this chapter, we present our algorithm called **ORAHWM**, which extends **ORAWM** to achieve the optimal rates for multirate multicast sessions. We present the general formulation of the multirate multicast resource allocation and provide a comparison between unirate and multirate multicast in allocating resources across multicast sessions.

6.1   Introduction

Heterogeneous multicast, or often called *multirate* multicast, is an efficient mode of data delivery for many, especially, real-time applications (e.g. teleconferencing, audio/video broadcasting). In multirate multicast, the receivers of a multicast group can receive service at different rates corresponding to their different characteristics. For example, receivers may receive different service rates, commensurate with their capabilities (e.g. processing power limitations) or based on their local network conditions (e.g. surrounding

---

[0]A version of this chapter has been submitted for publication [18, 19, 20]

wireless link states). Because of this receiver heterogeneity, multirate multicast allows the rate on some designated tree members to decrease in order to accommodate the congested receivers downstream, and hence it provides more flexibility in allocating rates across a multicast tree (see example in Section 6.2). Therefore, multirate multicast schemes have a great advantage over unirate multicast (homogeneous multicast) in adapting to diverse receiver requirements and heterogeneous network conditions.

The simplest way of attaining multirate multicast is by frame dropping. In this approach, intermediate nodes over the multicast tree may drop data frames to lower the rate for the downstream nodes. Another way is by hierarchical encoding or layered streaming which is particularly suitable for audio/video traffic. In this approach, the sender provides data in several layers organized in a hierarchy. Receivers subscribe to the layers cumulatively to provide progressive refinement [3]. This means that the receiver can only choose from a discrete set of data rates on each link[1]. Another method of attaining multirate multicast, which is particularly suitable for overlay multicast [7] is stream adaptation through transcoding [79] using some media gateways, which allows the receivers to choose its streaming rate on a continuous range. We assume that the network has any of these capabilities at least at some designated nodes (gateway nodes).

In this chapter, we present an optimal resource allocation algorithm for heterogeneous multicast over wireless ad hoc networks. In order to understand the advantage that multirate multicast promises over unirate multicast, we need to mention some draw-

---

[1]Refer to the techniques explained in Chapter 4 for how to calculate the transmission rates on each link in order to achieve optimal resource utilization

backs that our algorithm **ORAWM** (explained in previous chapter) suffers. One of these drawbacks is the inability of **ORAWM** to efficiently allocate network resources for multicast groups that have some congested group members (receivers). For such multicast groups, **ORAWM** tends to allocate rates based on the most congested receivers which is an overall waste of network resources. Another drawback is that **ORAWM** tends to discriminate against large multicast groups (i.e. multicast groups that go through more wireless contention domains and links) by giving them lower rates because large groups utilize more network resources. However, large multicast groups with large number of receivers should be favored over large groups with low number of receivers because the first would mean more data frames transmitted, and hence more network throughput. Indeed, this problem can be remedied by weighting the utility functions assigned to multicast groups giving large receiver groups more priority over low receiver groups. This can simply be negotiated when a receiver joins the group by sending the source the weight (or differentiation gain) of the utility function assigned to that receiver. The source will then use the aggregate weight of all receivers in maximizing the overall utility of the system, hence, favors large receiver groups over low receiver ones. Note that the transmission rate from a multicast group node needs to be equal to the maximum of the rates of all receivers downstream and is not allowed to exceed the transmission rate coming from upstream nodes (see Section 6.2.2 for more information).

The problem of resource allocation for multirate multicast has not been explored in the context of wireless ad hoc networks where resources are allocated for multicast subflows across multiple contention domains (i.e. see Chapter 5). Resource allocation for multirate multicast in wired networks has been studied in [44, 45]. An iterative algorithm based on subgradient techniques [80] has been employed to account for the non-differentiability of the primal problem. The authors in [7] proposed an overlay strategy for allocating resources over a multirate multicast tree by considering each link as a point-to-point unicast session. Rates are then allocated across each unicast session such that aggregate utility across all unicast sessions is maximized.

The remainder of this chapter is organized as follows. In Section 6.2, we explain new terminology used for heterogeneous multicast and formulate the optimization problem. The approach for multirate multicast is presented in Section 6.3. We present our distributed asynchronous algorithm for the heterogeneous case in Section 6.4. We provide the simulation results in Section 6.4.1. Finally, we conclude in Section 6.6.

6.2 Model and problem formulation

6.2.1 Model and notations

In addition to the network model explained in Section 5.2 we further divide the multicast tree nodes into *gateway nodes* and *relay nodes* as shown in Figure 6.1. Gateway nodes are the nodes that have rate control capabilities through one of the methods explained before, (e.g. layered transmission, transcoding, frame dropping, etc.). Relay nodes, on the other hand, merely forward data frames without performing any rate control functionality. v_i

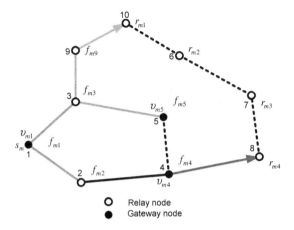

Figure 6.1: Multirate multicast network model.

denotes a gateway node at node i. If v_i is a member of multicast tree m (hence denoted as v_{mi}), then v_i can control the rate of the downstream nodes. F_v denotes the set of multicast subflows that belong to the multicast subtree starting from gateway node v and ending at either a terminal node or another gateway node. This subtree is denoted as T_v. $\Upsilon_m = \{v_{m1}, v_{m2}, ...\}$, is the set of all gateway nodes that are members of group m, and Υ is the set of all gateway nodes on all multicast trees $\forall m \in M$. Each multicast group has at least one gateway node (group source is considered a gateway node) to control the rate to the downstream nodes. We denote $\pi_m(v)$ to be the parent gateway node of gateway node v by going upstream towards the source of group m (e.g. $v_{m1} = \pi(v_{m4})$). Note that the source node has no parent gateway node (i.e. $\pi(v_{m1}) = \emptyset$). Also, note that for one multicast group, the rate that a gateway node is using for transmission at any given time must be greater than or equal to the maximum rate of all downstream gateway/receiver nodes. For example, in Figure 6.1, the rate used by v_{m1} for transmission must be greater than or equal to the maximum rate used by any of the gateway nodes v_{m4}, or v_{m5}. This adds a set of new constraints to the resource allocation problem which can be formulated by the following linear inequalities [2].

$$x_v \leq x_{\pi_m(v)} \quad \forall v \in \Upsilon_m \text{ s.t. } \pi_m(v) \neq \emptyset \quad \forall m \in M \qquad (6.1)$$

where x_v is the rate used by gateway node v and $x_{\pi_m(v)}$ is the rate used by parent gateway of gateway node v across the multicast group m.

Example

We present an example to illustrate the notations and highlight the main difference between unirate and multirate multicast in allocating rates in an ad hoc network. Figure 6.2 shows an example of an ad hoc network where there are eight nodes connected through wireless links as shown in the figure. The network contains two sessions m_1, and m_2 where m_2 has a traffic with fixed rate 800kbps as shown in the figure. m_1 uses Node 1 as the group source, and receiving Nodes 5, 6, whereas m_2 uses Node 7 as the group source and receiving Node 8. The aggregated subflows are represented by one node in the contention graph as shown in Figure 6.2(b). Assume that the channel capacity is 1Mbps, which means that the aggregate rate for each maximal clique cannot exceed 1Mbps. In this case, the rate on subflow f_{m4} cannot exceed 200Kbps because the traffic on that

---

[2]Here we formulate the constraints without using maximum functions to keep the constraints linear.

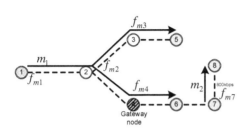 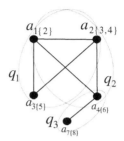

(a) An example of a multirate multicast ad hoc network   (b) Multicast contention graph

$$G_c = (V_c, E_c)$$

Figure 6.2: Example for resource allocation in unirate and multirate multicast.

subflow contends with f_{m7} and hence they both exist in the same maximal clique. Using unirate multicast, we cannot assign a rate to group m_1 more than 200Kbps because one of the receivers on this multicast is congested. This means that using unirate multicast we allocate the rate based on the most congested receiver. On the other hands, multirate multicast using one gateway node at Node 4 can make the rate allocation more efficient because, in this case, the rate used by source node 1 is allowed to exceed 200Kbps provided that gateway node 4 will adjust this rate to 200Kbps before forwarding the traffic to the downstream nodes. It can be shown that the rate used by source node 1 can be increased to 333Kbps in this case.

6.2.2   Mathematical Formulation

Here, we assign a utility function $U_v(x_v)$ for each gateway node on every multicast group $m \in M$ to measure the degree of satisfaction based on assigning a specific rate x_v to that gateway node. The optimization problem is to find the set of rates assigned to all gateway nodes for all multicast groups such that the aggregated utility function of all gateway nodes is maximized. This can be formulated with the following modified set of constraints:

$$\mathbf{P}' : \quad \textbf{maximize} \quad \sum_{v \in \Upsilon} U_v(x_v) \tag{6.2}$$

$$\textbf{subject to} \quad \sum_{v:(F_v \cap V_c^q) \neq \emptyset} x_v \Gamma_{qv} \leq c_q \quad \forall q \in Q \tag{6.3}$$

$$x_v \leq x_{\pi_m(v)} \quad \forall v \in \Upsilon_m \ \text{s.t.} \ \pi_m(v) \neq \emptyset \quad \forall m \in M$$

$$x_v \in I_v \quad \forall v \in \Upsilon$$

where Γ_{qv} represents the number of multicast subflows which belong to both clique q and the subtree T_v. With this problem formulation and the set of assumptions we stated in Section 5.2.2, we can leverage our solution approach discussed in Section 5.3 as we will see in the following section. Indeed, if we restrict each multicast group to have only one gateway (source) node, then the constraints in Equation 6.1 will be eliminated and problem $\mathbf{P}'$ will reduce to problem $\mathbf{P}$ discussed in Section 5.2.

6.3 Solution approach

Similar to what we did for problem P, we define the Lagrangian function, this time, $L(x, p, p')$ as follows:

$$
\begin{aligned}
L(x, p, p') &= \sum_{v \in \Upsilon} U_v(x_v) + \sum_{q \in Q} p_q(c_q - x_v \, \Gamma_{qv}) + \sum_{v \in \Upsilon} p'_v(x_{\pi_m(v)} - x_v) \\
&= \sum_{v \in \Upsilon} \left[U_v(x_v) - x_v \left(\sum_{q \in Q} p_q \Gamma_{qv} + p'_v - \sum_{v' \in \Lambda_m(v)} p'_{v'} \right) \right] + \sum_{q \in Q} p_q c_q \qquad (6.4)
\end{aligned}
$$

where $\Lambda_m(v)$ is the set of all children gateway nodes of node v (if any) along multicast tree m. Vectors $p = (p_q \; \forall q \in Q)$, and $p' = (p'_v \; \forall v \in \Upsilon)$ are two vectors of Lagrange multipliers. Γ_{qv} represents the number of multicast subflows that belong to subtree T_v and clique q simultaneously. Again we notice that the first term of Equation (6.4) is separable in x_v, and this entails

$$
\begin{aligned}
\max_{x_v \in I_v} \sum_{v \in \Upsilon} \left[U_v(x_v) - x_v \left(\sum_{q \in Q} p_q \Gamma_{qv} + p'_v - \sum_{v' \in \Lambda_m(v)} p'_{v'} \right) \right] \\
= \sum_{v \in \Upsilon} \max_{x_v \in I_v} \left[U_v(x_v) - x_v \left(\sum_{q \in Q} p_q \Gamma_{qv} + p'_v - \sum_{v' \in \Lambda_m(v)} p'_{v'} \right) \right]
\end{aligned}
$$

which means that this objective function can be divided into $|\Upsilon|$ separate subproblems. Each subproblem for subtree T_v can be solved locally if the values of clique prices $p_q \; \forall q :$ $(F_v \cap V_c^q) \neq \emptyset$, gateway forwarding price p'_v, and all children gateway forwarding prices $p'_{v'} \; \forall v' \in \Lambda_m(v)$ are known. The objective function of the dual problem then becomes:

$$
\begin{aligned}
D(p, p') &= \max_{x_v \in I_v} L(x, p, p') \\
&= \sum_{q \in Q} p_q c_q + \sum_{v \in \Upsilon} \max_{x_v \in I_v} \left[U_v(x_v) - x_v \left(\sum_{q \in Q} p_q \Gamma_{qv} + p'_v - \sum_{v' \in \Lambda_m(v)} p'_{v'} \right) \right]
\end{aligned}
$$

and the dual problem $\mathbf{D}'$ for the primal problem $\mathbf{P}'$ can then be defined as follows:

$$\mathbf{D}' : \min_{\substack{p \geq 0 \\ p' \geq 0}} D(p) \tag{6.5}$$

6.3.1 Interpretation of prices

Consider $P_v(T_v)$ as the profit of the subtree T_v which can be defined as follows:

$$P_v(T_v) = U_v(x_v) - x_v\left(\sum_{q \in Q} p_q \Gamma_{qv} + p'_v - \sum_{v' \in \Lambda_m(v)} p'_{v'}\right) \tag{6.6}$$

This profit represents the difference between the utility that subtree T_v gains by having rate x_v (i.e. $U_v(x_v)$) minus the summation of prices (denoted by $\tilde{U}(x_v)$) that this subtree has to pay for gaining such transmission rate, which is defined as

$$\tilde{U}(x_v) = \sum_{q \in Q} p_q \Gamma_{qv} x_v + p'_v x_v - \sum_{v' \in \Lambda_m(v)} p'_{v'} x_v \tag{6.7}$$

This summation of prices is divided into three components:

- $\sum_{q \in Q} p_q \Gamma_{qv} x_v$ which can be interpreted as the total price for utilizing resources on all cliques $\forall q \in Q$ such that $F_v \cap V_c^q \neq \emptyset$. In this case, p_q can be interpreted, as before, as the price per unit bandwidth consumed at clique q.

- $p'_v x_v$ is the price that subtree T_v must pay to the parent subtree of the same group in order to have traffic with rate x_v forwarded to it. In this case, p'_v is the price per unit bandwidth for forwarding traffic to subtree T_v.

- $\sum_{v' \in \Lambda_m(v)} p'_{v'} x_v$ is the total revenue that subtree T_v gains by forwarding traffic with rate x_v to all children subtrees with each term $p'_{v'} x_v$ indicating the revenue for forwarding traffic to subtree $T_{v'}$ such that $v' \in \Lambda_m(v)$.

Note that at optimality, $p'_v = 0$ if $x_v < x_{\pi_m(v)}$ since p'_v indicates the price when the constraints in Equation (6.1) are violated or the maximum possible rate is used (i.e. $x_v = x_{\pi_m(v)}$). This means that a subtree $T_{v'}$ is not charged for using rate x_v if this rate is *less than* the rate at parent gateway node $\pi_m(v)$.

For p_q we can, similarly, define the price for one subflow $f_{vi} \in T_v$ as the total price for consuming bandwidth on all maximal cliques $q \in Q$ as follows:

$$\lambda_{vi} = \sum_{q:(f_{vi} \in V_c^q) \neq \emptyset} p_q \tag{6.8}$$

and we can also define the aggregated price for subtree T_v as a result of consuming bandwidth on all maximal cliques $q \in Q$ as follows:

$$\lambda_v = \sum_{q:(F_v \cap V_c^q) \neq \emptyset} p_q \Gamma_{qv} \tag{6.9}$$

Equation (6.9) calls for a technique to calculate this price in a distributed network environment using the accumulated subtree price $\lambda_v(i)$ similar to the one explained in Section 5.4.1. This time $\lambda_v(i)$ can be defined as follows

$$\lambda_v(i) = \frac{\lambda_v(\pi_m(i)) + \lambda_{v\pi_m(i)}}{K_{v\pi_m(i)}} \quad \forall i \in T_v \tag{6.10}$$

where $K_{v\pi_m(i)}$ is the cardinality of subflow $f_{v\pi_m(i)}$.

6.3.2 Aggregated subtree price calculation

In Section 6.4 we will explain the iterative method for calculating both clique price p_q (hence subflow price from Equation (6.8)), and the forwarding price for each gateway node p'_v. In order to calculate the total price defined by Equation (6.7) at any gateway node v, we need to calculate the accumulated price on each branch recursively using Equation (6.10) until we hit either a terminal node or another gateway node $v' \in \Lambda_m(v)$. Each gateway node $v' \in \Lambda_m(v)$ subtracts the forwarding price $p'_{v'}$ from the accumulated price to get the net price for the branch leading to that gateway node. Children gateway and terminal nodes which are part of T_v then send the net price value back to node v to calculate the subtree aggregate price per unit bandwidth $\lambda(T_v)$ by simply aggregating all net branch prices and the forwarding price p'_v as follows:

$$\lambda(T_v) = \lambda_v + p'_v - \sum_{v' \in \Lambda_m(v)} p'_{v'} \tag{6.11}$$

6.4 Optimal Resource Allocation for Heterogeneous Wireless Multicast (ORAHWM)

To solve the primal problem **P** in a distributed asynchronous network environment, we use the gradient projection method as explained in Section 5.4. The clique prices $p_q(t+1) \; \forall q \in Q$ and forwarding prices $p'_v \; \forall v \in \Upsilon$ are calculated iteratively as follows:

$$p_q(t + 1) = [p_q(t) - \alpha \frac{\partial D(p(t))}{\partial p_q}]^+ \tag{6.12}$$

$$p'_v(t + 1) = [p'_v(t) - \alpha \frac{\partial D(p'(t))}{\partial p_v}]^+ \tag{6.13}$$

where $\alpha > 0$ is the gradient step size. Since $U_v \; \forall v \in \Upsilon$ are concave functions, $D(p, p')$ is continuously differentiable and the gradients for $D(p, p')$ with respect to p, and p' are defined as follows:

$$\frac{\partial D(p, p')}{\partial p_q} = c_q - \sum_{v:(F_v \cap V_c^q) \neq \emptyset} x_v(t)\Gamma_{qv} \qquad q \in Q \tag{6.14}$$

$$\frac{\partial D(p, p')}{\partial p'_v} = x_{\pi_m(v)}(t) - x_v(t) \qquad v, \ \pi_m(v) \in V_m \tag{6.15}$$

Substituting in Equations (6.12), and (6.13) we get the supply and demand equations

for calculating p, and p' as follows:

$$p_q(t+1) = [p_q(t) - \alpha(c_q - \sum_{v:(F_v \cap V_c^q) \neq \emptyset} x_v(t)\Gamma_{qv})]^+ \tag{6.16}$$

$$p'_v(t+1) = [p'_v(t) - \alpha(x_{\pi_m(v)}(t) - x_v(t))]^+ \tag{6.17}$$

We calculate the subtree aggregate price $\lambda(T_v, t+1)$ defined by Equation (6.11) at

time $(t+1)$ using the clique, and forwarding price values from Equation (6.16), and (6.17)

as explained in Section 6.3.2. Finally, the transmission rate used by gateway node v at

time $(t+1)$ is calculated as follows:

$$x_v(t+1) = [U'_v(\lambda(T_v, t+1))]^{W_v}_{w_v} \tag{6.18}$$

The details for the asynchronous algorithm for heterogeneous wireless multicast is

shown in Figure 6.3. In order to understand the association of this algorithm with

the network architecture, we assume that each node i in the network has zero or more

multicast subflows $f_{vi} \ \forall v \in \Upsilon$ depending on the traffic passing by this node. Even

though, the algorithm suggests that the *clique procedure* at clique q can be performed by

a designated node from that clique (i.e. clique master), in our simulations we perform

the clique procedure at each node i separately for all cliques that have $f_{vi} \cap V_c^q \neq \emptyset$. The

Clique procedure (by clique q): At times $t \in T_q$

1. Receive rates $x_v(t')$ from all subtrees T_v where $F_v \cap V_c^q \neq \emptyset$

2. Update clique price as follows
$$p_q(t+1) = [p_q(t) - \alpha(c_q - \sum_{v:(F_v \cap V_c^q) \neq \emptyset} \hat{x}_v(t)\Gamma_{qv})]^+, \quad \forall q \in Q$$

3. Send $p_q(t+1)$ to all nodes of group m such that $F_m \cap V_c^q \neq \emptyset$

Subflow procedure (by subflow f_{vi}): At times $t \in T_m^i$

1. Receive prices $p_q(t')$ from all maximal cliques q where $f_{vi} \cap V_c^q \neq \emptyset$

2. Calculate the subflow price (per hop price) λ_{vi} as follows
$$\lambda_{vi}(t+1) = \sum_{q:(f_{vi} \cap V_c^q) \neq \emptyset} \hat{p}_q(t)$$

3. Calculate the accumulated price on each branch $b_{vij} \in f_{vi}$
$$\lambda_v(j, t+1) = \frac{\lambda_v(i,t) + \lambda_{vi}(t+1)}{K_{vi}}$$

4. Forward $\lambda_v(j, t+1)$ to all children subflows of f_{vi}, if no children, send $\lambda_v(j, t+1)$ to v.

Subtree procedure (by gateway v): At times $t \in T_v$

1. Receive the net prices $\lambda_v(i, t') - p_i'(t')$ from all terminal nodes of T_v ($i.e. \forall i \in \Im_v$), and all children gateway nodes $\forall i \in \Lambda_m(v)$ //(note: $p_i' = 0 \ \forall i \in \Im_v$).

2. If $\pi_m(v) \neq \emptyset$ //(i.e. $v \neq s_m$)
   - Receive rate $x_{\pi_m(v)}(t'')$ from parent subtree of T_v
   - Calculate the next forwarding price $p_v'(t+1)$ as follows:
   $$p_v'(t+1) = [p_v'(t) - \alpha(\hat{x}_{\pi_m(v)}(t) - x_v(t))]^+$$
   Else $p_v'(t+1) = 0$

3. Calculate the next subtree aggregate price $\lambda(T_v, t+1)$ as follows:
$$\lambda(T_v, t+1) = p_v'(t+1) + \sum_{i \in (\Im_m \cup \Lambda_m(v))} (\hat{\lambda}_v(i,t) - \hat{p}_i'(t))$$

4. Calculate the next subtree rate as follows:
$$x_v(t+1) = U_v'^{-1}(\lambda(T_v, t+1))$$

5. Send $x_v(t+1)$ to all cliques q where $F_v \cap V_c^q \neq \emptyset$

Figure 6.3: ORAHWM: Asynchronous distributed algorithm.

subflow procedure is performed by each node i that has one or more multicast subflows f_{vi} $\forall v \in \Upsilon$ by simply calculating the accumulated prices at the branches of f_{vi} based on the accumulated price at i. Finally, *active gateway nodes* [3] perform the *subtree procedure* by calculating the optimal rate $x_v(t+1)$ based on the aggregated prices for this subtree.

The estimation of the price and rate values at time t from the received values at time instances in the range $(t - B) \leq t' \leq t$ may follow any update policy such as the *latest instant update* or the *latest average update* as explained in Section 5.4.3. For example, the estimation of $\hat{x}_v(t)$ at time t from the values of $x_v(t')$ at times $(t-B) \leq t' \leq t$ follows the general weighted average model

$$\hat{x}_v(t) = \sum_{t'=t-B}^{t} b_i^q(t', t)\, x_v(t') \quad \text{where} \sum_{t'=t-B}^{t} b_i^q(t', t) = 1$$

Using this model, we proved in Chapter 5 through analytical modeling and simulations the convergence of the asynchronous algorithm regardless of the update policy used which was a key feature for these set of algorithms. The following experiments also show the same results for the case of heterogeneous wireless multicast.

6.4.1   Simulation results

As in Chapter 5, we use the utility functions $U_v(x_v) = g_v \ln(x_v)$ $x_v > 0$ for imposing proportional fairness amongst the multicast groups where g_v is the differentiation gain for gateway node v, i.e. $x_v(t) = g_v/\lambda_v(t)$.

---

[3]Gateway nodes that have traffic from one or more multicast groups passing by them.

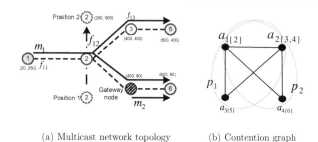

(a) Multicast network topology   (b) Contention graph

Figure 6.4: Effect of changing differentiation gains on the calculated rates and aggregate

utility.

Effect of changing differentiation gains on the calculated rates and aggregate utility

In this experiments, we study the effect of changing the differentiation gains on the calculated rates for unirate and multirate multicast sessions. We consider the small topology shown in Figure 6.4. Two sessions m_1, and m_2 are sharing this network with source and receiver nodes as shown in the figure. We consider three cases where we change the differentiation gain and show the effect on the calculated rates in each case. Case 1 is the unirate multicast where we use one gateway/source node for each multicast group, and we use equal differentiation gains for both sessions (i.e. $g_{v_{11}} = g_{v_{24}} = 5$). For both cases 2 and 3, m_1 uses gateway node 4 for rate control . Case 2 uses differentiation gains $g_{v_{11}} = 3, g_{v_{14}} = 2$ whereas case 3 uses $g_{v_{11}} = 4, g_{v_{14}} = 1$. In all cases we start both multicast groups at $t = 20$ seconds, we fix all other parameters including $\alpha = 3 \times 10^{-7}$ and we set the channel capacity for all maximal cliques to 1Mbps.

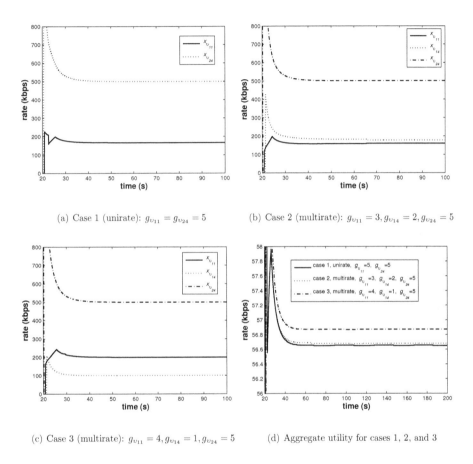

(a) Case 1 (unirate): $g_{v_{11}} = g_{v_{24}} = 5$

(b) Case 2 (multirate): $g_{v_{11}} = 3, g_{v_{14}} = 2, g_{v_{24}} = 5$

(c) Case 3 (multirate): $g_{v_{11}} = 4, g_{v_{14}} = 1, g_{v_{24}} = 5$

(d) Aggregate utility for cases 1, 2, and 3

Figure 6.5: Effect of changing differentiation gains on the calculated rates and aggregate utility.

Figure 6.5 shows the calculated rates and the aggregated utility for the three cases. We notice that for Case 1 (unirate), as expected, our algorithm **ORAHWM** will discriminate against session m_1 because it uses more wireless links and hence utilizes more network resources. This happens because, for unirate, **ORAHWM** deals with each session as one entity regardless of how large this session is and how many links it uses. Multirate with additional gateway nodes can reduce this effect by providing more flexibility to assign more priority to some parts of the tree which in turn affects the aggregate utility of the entire system. This is depicted by the results in Figures 6.5(b), and 6.5(c). We notice that by increasing the differentiation gain for $T_{v_{11}}$, we can increase the aggregate utility (shown in Figure 6.5(d)).

Effect of dynamic capacity and mobility on the convergence of ORAHWM

In this experiments, we study the effect of changing network conditions including changing capacity and node mobility on the convergence of our algorithm **ORAHWM**. We consider the same topology and multicast sessions shown in Figure 6.4 and we use the $g_{v_{11}} = 4, g_{v_{14}} = 1$.

First, we study the effect of measuring the real capacity on each clique using the MAC layer channel capacity estimator as explained in Section 5.5.2. Figure 6.6 shows the result of using a time-varying channel capacity realized by the MAC scheduler IEEE 802.11 DCF. From this figure, we observe that although the achieved channel capacity changes with time, our algorithm continuously tracks the change in channel capacity fairly well and provides proportional fairness amongst all the multicast sessions based on the current available channel capacity.

We also study the impact of mobility and route changes on the convergence of our algorithm by generating a mobility pattern where Node 2 moves from *Position 1* to *Position 2* as shown in figure with average speed of 3 m/s and pause time of 20 seconds. Figure 6.7 shows the rates calculated by our algorithm with time. The figure shows three different regions depending on the change of routes resulting from the node mobility. In Region 1, only Node 6 is receiving traffic for both multicast sessions. As expected in this case, our algorithm converges to the same rates of Case 3 in the previous experiment. As Node 2 moves to Region 2, the routes for which are shown in Figure 6.4, both receivers at Nodes 5 and 6 become active for Session m_1 and the optimal rates converge to the same values, after some transient period, despite the route changes. When Node 2 moves to Region 3, both the receiver at Node 6 and Node 4 become inactive for Session m_1, and Session m_2 can now use the whole channel for its traffic. Therefore, the optimal rate for m_2 in this case is 1Mbps whereas the capacity is divided amongst the three subflows f_{11}, f_{12}, and f_{13} for m_1.

Effect of using multirate on the total throughput for multicast flows

In this experiment we study the effect of using gateway nodes for rate control as part of a multicast group. Consider Figure 6.8 which shows two multicast groups m_1 and m_2 sharing an ad hoc network on 11-nodes as shown in the figure. m_1 uses gateway/source Node 1 (i.e. v_{11}), and has 3 receivers, namely: r_7, r_8, and r_9, whereas m_2 uses gateway/source Node 6 (i.e. v_{26}) and has two receivers, namely: r_{10}, and r_{11}. Here, to study the impact

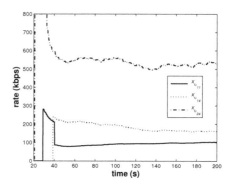

Figure 6.6: Calculated rates with dynamic capacity: $g_{v_{11}} = 4, g_{v_{14}} = 1, g_{v_{24}} = 5$.

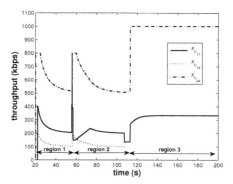

Figure 6.7: Calculated rates with mobility: $g_{v_{11}} = 4, g_{v_{14}} = 1, g_{v_{24}} = 5$.

of using multirate multicast we consider two cases. Case 1 is the unirate multicast with equal differentiation gains for both multicast groups (i.e. $g_{v_{11}} = g_{v_{26}} = 3$). For Case 2, m_1 uses an additional gateway node at 4 (i.e. v_{14}) for rate control. In this case, we set $g_{v_{11}} = 2$, $g_{v_{14}} = 1$ so the total differentiation gain is similar to Case 1, and we set

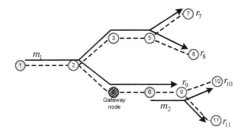

Figure 6.8: Multirate multicast network topology.

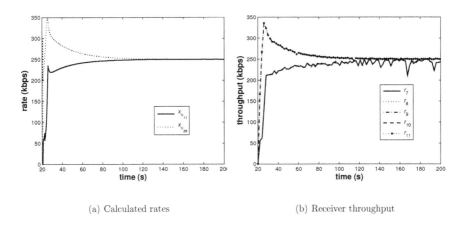

(a) Calculated rates (b) Receiver throughput

Figure 6.9: Case 1 (unirate): calculated rate and throughput without using gateway node v_{14}.

Figures 6.9 and 6.10 show the calculated rates and receiver throughput for Cases 1 and 2, respectively. We notice that in each case convergence is attained, and the throughput achieved by all receivers on each group tracks the calculated rates appropriately. Com-

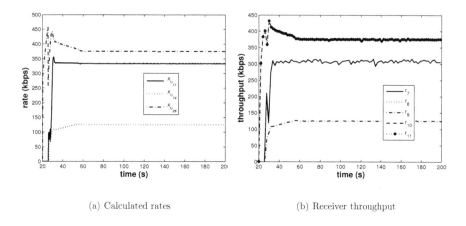

(a) Calculated rates   (b) Receiver throughput

Figure 6.10: Case 2(multirate): calculated rate and throughput using gateway node v_{14}

for rate control on m_1.

paring the two figures, we notice the effect of using gateway node v_{14} for m_1 which lowers

the optimal rate on the subtree $T_{v_{14}}$ (i.e. $x_{v_{14}}$) allowing the other rates $x_{v_{11}}$ and $x_{v_{26}}$

to increase drastically. This happens because we set the differentiation gain $g_{v_{14}} = 1$

giving this subtree lower priority based on our knowledge that this subtree has only one

receiver (r_9) and the surrounding area has traffic load more than for example $T_{v_{11}}$ and

we used v_{14} to give us the flexibility of setting $x_{v_{14}}$ accordingly. Such knowledge can

either be communicated between the receivers and gateway nodes or tuned manually by

an administrator.

To study the effect of this heterogeneity within m_1, we measure the aggregate utility

and the total throughput achieved by each group for Cases 1 and 2. Figures 6.11 and 6.12

show the results for these measurements. We see from Figure 6.11 that the aggregate utility achieved for Case 2 is better as a result of using gateway node v_{14} because both rates $x_{v_{11}}$ and $x_{v_{26}}$ increased significantly by reducing $x_{v_{14}}$. This increase in rates caused the overall throughput achieved by both multicast groups to increase drastically (i.e. $\approx$ 30%) as shown in Figure 6.12.

6.5 Related Work

In this section, we evaluate and compare qualitatively the contributions in Chapters 5 and 6 in light of the previous work of resource allocation in wired and wireless networks.

The problem of optimal and fair resource allocation has been widely studied in the context of wired networks. Among these studies (e.g. [10], [1], [7], [45], [44]), price-based methods have shown to be effective in achieving a decentralized solution for rate allocation. Resource allocation for unicast sessions has been covered in [10], [1], whereas [7], [45], [44] discussed the problem for multicast sessions. Although the role of the price in achieving a decentralized solution in our model is analogous to these studies, there are fundamental issues that our solution addresses and they have not been addressed by these models. Some of these issues can be highlighted as follows:

- The location-based contention and the shared wireless channel characteristics of the wireless ad hoc networks represent major challenges in our model and require unique treatment in allocating network resources. Our model addresses these spe-

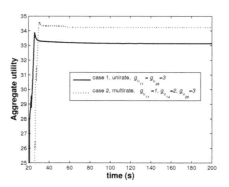

Figure 6.11: Aggregate utilities for Cases 1 (unirate) and 2 (multirate).

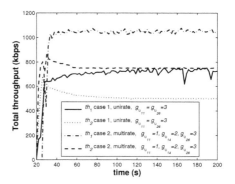

Figure 6.12: Total throughput for each multicast group for Cases 1 and 2: th_1 is total throughput for m_1 and th_2 is total throughput for m_2.

cial characteristics by employing the protocol model in constructing contention domains and maximal cliques that represent network resource entities.

• The static nature of the end-to-end sessions in wired networks makes the convergence in wired networks simpler and more tractable. On the other hand, as part of our solution we addressed the behavior of the decentralized algorithm as a result of online changing conditions in the network such as mobility, route changes, and time-varying channel capacity.

- None of these studies addressed the deployment issues as a result of deploying these solutions in real networks. One of the major contributions that we have emphasized is the design of a cross-layer architecture for realizing optimal fair resource allocation in real ad hoc networks. Such architecture adds more practicality to our results and demonstrates more potential to our solution.

Resource allocation for single-hop and unicast flows has been studied in wireless ad hoc networks. Resource allocation using MAC-layer fair scheduling for single-hop flows has been studied in [46, 47, 48]. Such techniques, however, do not provide end-to-end resource optimization and therefore fairness among the multihop flows cannot be achieved only by local MAC layer scheduling. In comparison, we address the end-to-end multihop flows, and provide a solution that takes into consideration the network topology and the coordination between different hops of the end-to-end flows in enhancing the overall network's resource utilization. In doing that, our solution uses the clique price as a signal to coordinate the relationship between the different hops and regions to attain global resource allocation. The end-to-end resource allocation for unicast flows has been covered in [6, 49] using clique price for coordinating end-to-end resource allocation. We have provided a quantitative comparison between multicast and unicast in allocating resources across end-to-end sessions in Section 5.2.2. This comparison shows the major advantage of supporting wireless multicast in increasing the network's available bandwidth and reducing the number of end-to-end sessions for one-to-many communication. This nominates wireless multicast as a very attractive scheme, especially for multimedia applications. Also, it should be noted that our model represents the general formulation of resource allocation in wireless networks and can be reduced to the unicast model if all the end-to-end sessions are unicast. Therefore, our solution has major differences and additional characteristics both in the problem formulation and the architecture for deploying the algorithm in ad hoc networks, which can be highlighted as follows:

- Our model promotes using the one-hop broadcast feature of the wireless medium in multiplying the network's available bandwidth through the notion of multicast subflows. Such representation provides flexibility in allocating rates across the different part of the multicast tree. It also allows supporting multirate multicast which provides ultimate flexibility in allocating rates while increasing network's resource utilization.

- Dividing end-to-end sessions into multicast subflows, however, introduces major challenge in our model that requires unique treatment especially for constructing contention domains and maximal cliques (see Section 5.2.1).

- Calculating the price for multicast sessions in a decentralized way represents another challenge in our model, which motivated the development of our group price calculation mechanism described in Section 5.4.1.

- The design of the network architecture for resource allocation has been influenced significantly by supporting multicast sessions. This appears in integrating multicast aware routing protocol (MAODV) and group HELLO packets for conveying aggregate subflow status in the routing layer. We have suggested the use of Multicast-aware MAC layer protocol (MMP) for reliable multicast transfer, and multicast aware measurement-based technique for channel capacity estimation in the MAC layer. Lastly, we have suggested modifying the transport layer to deal with multiple feedback packets as part of the group procedure to calculate the group price.

- To demonstrate the versatility of our solution for different network environments, we have considered a series of implementations based on the information distribution and channel dynamics and showed the convergence behavior in all these environments.

6.6 Concluding Remarks

In this chapter we have presented the resource optimization algorithm for the case of multirate multicast (**ORAHWM**) over multihop ad hoc networks. We have introduced the notion of gateway nodes used to control the rates for multirate multicast groups and provided the optimization model that realizes the optimal rates used by each gateway node in order to maximize the overall aggregate utility for the entire system. Utilizing the flexibility of using gateway nodes across the multicast trees, **ORAHWM** is expected to increase the aggregate utility of the system and boost the overall throughput achieved by each multicast group provided that the differentiation gains are set appropriately.

Chapter 7

Conclusions and Future Research

We conclude this thesis with a summary of our contributions and suggestions for future work.

7.1 Summary

The main objective of this thesis was to offer a set of viable solutions that can close some essential gaps in providing optimal resource allocation in wired and wireless networks. We motivated the problems presented in this thesis and provided survey of the latest techniques in the area of resource allocation in Chapter 2. Later chapters then presented our major contributions which are summarized as follows:

Chapter 3 presented a group of algorithms for calculating the optimal classification for a set of traffic streams with diverse QoS requirements for a link model with a predetermined service levels or predetermined class weights. It also studied the effect of the three proposed service differentiation (γ) estimation methods on the quantization overhead and provided a comparison study based on complexity analysis and simulation. Using the dropping probability as the QoS metric, a conclusion has been deduced that using 4 or 5 service levels will achieve the compromise between complexity and minimizing the quantization overhead. Furthermore, the expected quantization overhead for stochastic QoS requirements has also been studied. Convergence analysis has then been provided to study the effect of sampling the QoS pdf functions on the quantization overhead. Results from this work appear in [12] [13] [14].

Chapter 4 presented the algorithms for calculating the optimal partitioning and bandwidth allocation for a set of traffic streams with diverse QoS requirements for a link model with variable service levels. Such algorithms can be useful for service providers to design the network service levels that achieve the best granularity based on the different distributions of the QoS requirements and bandwidth demands. It has been shown again that using 4 or 5 service levels will achieve the compromise between complexity and minimizing the quantization overhead using dropping probability as the QoS metric. Moreover, it has also been shown that the optimal QoS partitioning with bandwidth allocation (**OQP-OBA**) algorithm will guarantee graceful service degradation when the link is overloaded due to the absence of admission control. This work appears in [17].

Chapter 5 presented a new multicast-based algorithm and analytical model for resource optimization over multihop ad hoc networks (**ORAWM**). This algorithm is used to control the rates of the multicast sessions in such a way that guarantees the optimal resource utilization of the wireless network resources while achieving fairness amongst the multicast sessions utilizing the bandwidth-efficiency feature of multicast which increases the overall network available bandwidth. Based on using maximal cliques as the main resource entity in the network, the multicast sessions has been modeled in a form of contention domains, and the group price for the end-to-end multicast session has been calculated. This chapter also proposed a mechanism for calculating the group price based on the branch accumulated price which allows the calculation to occur in a totally distributed and asynchronous way. It has been shown through simulations that convergence can still be attained with slow network changes including channel dynamics and node mobility. This work appears in [18].

Chapter 6 presented the resource optimization algorithm for the case of multirate multicast (**ORAHWM**) over multihop ad hoc networks. It introduced the notion of gateway nodes used to control the rates for multirate multicast groups and provided the optimization model that realizes the optimal rates used by each gateway node in order to maximize the overall aggregate utility for the entire system. Utilizing the flexibility of using gateway nodes across the multicast trees, **ORAHWM** is expected to increase the aggregate utility of the system and boost the overall throughput achieved by each multicast group compared to **ORAWM** provided that the differentiation gains are set appropriately.

7.2   Future Work

In the following, we provide a list of possible future directions for research related to the work reported in this thesis.

1. QoS classification and partitioning techniques discussed in Chapter 3 and Chapter 4 have been evaluated using statistical simulation models. Studying the effect of these techniques in a topological network could be of practical significance. In other words, it would be interesting to simulate these proposed techniques in a network with a specific topology and study the effect of using such techniques on the overall resource utilization of the network.

2. Mapping the service levels discussed for QoS classification/partitioning to protocol entities such as labels in Multi Protocol Label Switching (MPLS) [81] will allow us to devise a signaling mechanism that can be used to negotiate the calculated service levels among the network Label Switched Routers (LSRs). Such mapping will be crucial to evaluate the performance of these techniques in MPLS-enabled network.

3. Although our asynchronous algorithms **ORAWM** and **ORAHWM** presented in Chapter 5 and Chapter 6 converge online to the optimal rates, the convergence behavior is critically sensitive to the value of the step size α. Therefore, it might be useful to design a technique which guarantees the convergence with less sensitivity on the step size. From the theory of optimization, several techniques (e.g. variations of Newton's method) exist for exactly this purpose [73] and they sometimes offer faster convergence compared to gradient methods. However, none of these methods can be implemented in a distributed environment with asynchronous computations. Therefore, designing a technique which achieves faster convergence with less sensitivity on α without sacrificing the important feature of distributed computations will have a

significant potential.

4. Extending the multicast-based models in Chapter 5 and Chapter 6 to capture some wireless physical characteristics like power may be crucial for supporting power control mechanisms for multihop wireless networks with multicast-based communication. Such a model will optimize some physical layer related objectives (e.g. signal to interference noise ratio (SIR)) subject to some power limitations on the network nodes.

5. Our multicast-based model in Chapter 6 addresses the heterogeneity in the multicast receivers in terms of rate requirements while assuming that all nodes relay the traffic to other nodes homogeneously. However, another significant aspect of heterogeneity might be in the physical characteristics of each node (e.g. battery power). Such heterogeneity will affect the willingness of each node to relay the traffic for other nodes with the ad hoc network. An additional cost parameter may be used in this case to measure the willingness of each node to relay traffic to other nodes. The objective then is to optimize the aggregate utility of all sessions after deducting the cost of relaying the traffic from one node to another across each session.

6. Mapping the protocol entities for implementing **ORAWM** and **ORAHWM** to one of the existing standards for multihop wireless networks (e.g. 802.15.4/ZigBee) could also be a potential area of research. Such mapping will allow us to study the effect of these multicast-based techniques on enhancing the functionality of a ZigBee network which would be of practical significance especially to SPs.

Bibliography

[1] R. J. La and V. Anantharam. Utility-based rate control in the internet for elastic traffic. *IEEE/ACM Transactions on Networking*, 10(2):272–286, April 2002.

[2] Y. Kogan W. Berger. Dimentioning bandwidth for elastic traffic in high-speed data networks. *IEEE/ACM Transactions on networking*, 8(5), October 2000.

[3] S. Paul X. Li and M. Ammar. Layered video multicast with retransmission (LVMR): Evaluation of hierarchical rate control. *In proceedings of IEEE Infocom*, March 1998.

[4] Y. Richard Yang, Min Sik Kim, and Simon S. Lam. Optimal partitioning of multicast receivers. *In proceedings of the 8th IEEE ICNP*, 2000.

[5] K. Gopalan and T. Chiueh. Multi-resource allocation and scheduling for periodic soft real-time applications. *In proceedings of Multimedia Computing and Networking, San Jose, CA, USA*, pages 34–45, January 2002.

[6] Y. Xue, B. Li, and K. Nahrstedt. Optimal resource allocation in wireless ad hoc networks: A price-based approach. *in Technical Report UIUCDCS-R-2004-2505, University of Illinois at Urbana-Champaign*, June 2004.

[7] Yi Cui, Y. Xue, and K. Nahrstedt. Optimal resource allocation in overlay multicast,. *In proceedings of the 11th International Conference on Network Protocols (ICNP), Atlanta, Georgia,* November 2003.

[8] J. Kuri P. Chaporkar. A network architecture for providing per flow delay guarantees with scalable core. *Journal for High Speed Networks,* 12(3):87–109, Jan 2002.

[9] A. K. Parekh and R. G. Gallager. A generalized processor sharing approach to flow control in integrated service networks: The single node case. *IEEE/ACM Transactions on Networking,* 1(3):344–357, June 1993.

[10] S. H. Low and D. E. Lapsley. Optimization flow control: Basic algorithm and convergence,. *IEEE/ACM Transactions on Networking,* 7:861–874, 1999.

[11] A. K. Parekh and R. G. Gallager. A generalized processor sharing approach to flow control in integrated service networks: The multiple node case. *IEEE/ACM Transactions on Networking,* 2(2):137–150, 1994.

[12] A. Mohamed and H. Alnuweiri. Optimal QoS-based classification for multi-class link models with predetermined service levels. *Submitted to IEEE/ACM Transactions on Networking,* March 2006.

[13] A. Mohamed and H. Alnuweiri. Dynamic programming QoS-based classification for links with limited service levels. *In proceedings of IEEE Conference on local area networks LCN, Sydney, Australia,* pages 51–58, November 2005.

[14] A. Mohamed and H. Alnuweiri. Optimal QoS-based classification for link models with predetermined service levels. *In proceedings of the 8th International Conference on Telecommunications, Contel, Zagreb, Croatia*, 2:375–382, June 2005.

[15] A. Mohamed and H. Alnuweiri. Dynamic programming QoS-based classification for links with limited service levels. *International Transactions on Computer Science and Engineering (GESTS)*, 19(1):97–108, October 2005.

[16] A. Mohamed and H. Alnuweiri. Stochastic QoS-based classification for link models with calculated service levels. *In proceedings of Conference on Communications, Computers and signal Processing, PACRIM*, pages 364–367, August 2005.

[17] A. Mohamed and H. Alnuweiri. QoS-based partitioning and resource allocation for link models with variable service levels. *Accepted for publication in IEEE ISCC, Sardinia, Italy*, June 2006.

[18] A. Mohamed and H. Alnuweiri. A distributed iterative algorithm for optimal rate allocation for homogeneous wireless multicast. *Accepted for publication in IST mobile & wireless communications summit, Greece*, june 2006.

[19] A. Mohamed and H. Alnuweiri. Optimal resource allocation for homogeneous wireless multicast. *Submitted to IEEE Globecom, San Francisco*, November 2006.

[20] A. Mohamed and H. Alnuweiri. Cross-layer distributed approach for optimal rate allocation for homogeneous wireless multicast. *Submitted to IEE Proceedings Commu-*

nications, special Issue on Wireless Mobile Networks: Cross-Layer Communication, April 2007.

[21] A. Mohamed and H. Alnuweiri. Design of a cross-layer optimization framework for rate allocation in wireless multicast. *Submitted to IEEE International conference on Mobile Ad-hoc and Sensor Systems (MASS), Vancouver, Canada,* October 2006.

[22] A. Mohamed and H. Alnuweiri. Cross-layer optimal resource allocation for heterogeneous wireless multicast. *Submitted to IEEE Journal on Selected Areas in Communications, special issue on cross layer optimized wireless multimedia communications,* 2007.

[23] A. Kaheel, T. Khattab, A. Mohamed, and H. Alnuweiri. Quality-of-service mechanisms in IP-over-WDM networks. *IEEE Communications Magazine,* 40(12):38–44, December 2002.

[24] X. Wang and K. Kar. Cross-layer rate control for end-to-end proportional fairness in wireless networks with random access. *In proceedings of the 6th ACM international symposium on Mobile ad hoc networking and computing, Urbana-Champaign, IL, USA,* pages 157–168, 2005.

[25] L. Tassiulas K. Kar, S. Sarkar. Achieving proportional fairness using local information in Aloha networks. *In IEEE Transactions on Automatic Control,* 49:1858–1863, November 2004.

[26] D. Bertsekas and R. Gallager. *Data Networks.* Prentice Hall, 1987.

[27] Z. Fang and B. Bensaou. Fair bandwidth sharing algorithms based on game theory frameworks for wireless ad-hoc networks. *In proceedings of the IEEE Infocom*, 2004.

[28] A. Elwalid and D. Mitra. Design of generalized processor sharing schedulers which statistically multiplex heterogeneous qos classes. *In proceedings of IEEE INFOCOM*, pages 1220–1230, 1999.

[29] Q. Ni, L. Romdhani, and T. Turletti. A survey of QoS enhancements for IEEE 802.11 wireless lan. *Journal of Wireless Communications and Mobile Computing, Wiley*, 4(5):547–566, 2004.

[30] X. Long Huang and B. Bensaou. On max-min fairness and scheduling in wireless ad-hoc networks: analytical framework and implementation. *In proceedings of IEEE/ACM MobiHoc, Long Beach, CA*, pages 221–231, October 2001.

[31] A. Legout, J. Nonnenmacher, and E. W. Biersack. Bandwidth allocation policies for unicast and multicast flows. In *Proceedings of IEEE INFOCOM'99*, pages 254–261, New York, NY, USA, 1999.

[32] H. C. Cankaya, S. Charcranoon, and T. S. El-Bawab. A preemptive scheduling technique for OBS networks with service differentiation. *In proceedings of Globecom*, pages 2704–2708, 2003.

[33] N. Christin, J. Liebeherr, and T. Abdelzaher. A quantitative assured forwarding service. *In proceedings of IEEE INFOCOM, New York*, 2:864–873, June 2002.

[34] Y. Chen, M. Hamdi, and D. H. K. Tsang. Proportional QoS over OBS networks. *In proceedings of IEEE Globecom, New York*, pages 1510–1514, 2001.

[35] S. Blake, D. Black, M. Carlson, E. Davies, Z.Wang, and W.Weiss. An architecture for differentiated services. *IETF RFC 2475*, December 1998.

[36] G. Bianchi, N. Blefari-Melazzi, and M. Femminella. Per-flow QoS support over a stateless differentiated services IP domain. *The International Journal of Computer and Telecommunications Networking*, 40(1):73–87, 2002.

[37] M. Yang, Y. Huang, J. Kim, M. Lee, T. Suda, and M. Daisuke. An end-to-end QoS framework with on-demand bandwidth reconfiguration. *IEEE INFOCOM*, March 2004.

[38] K. Takagaki, H. Ohsaki, and M. Murata. Analysis of a window-based flow control mechanism based on TCP vegas in heterogeneous network environment. *In proceedings of IEEE International Conference on Communications ICC*, 10:3224–3228, June 2001.

[39] H. Ohsaki, M. Murata, T. Ushio, and H. Miyahara. A control theoretical analysis of window-based flow control mechanism based on TCP vegas,. *High Quality Internet Workshop*, October 1998.

[40] D. H. Lorenz and A. Orda. Optimal partition of QoS requirements on unicast paths and multicast trees. *IEEE/ACM Transactions on Networking*, 10(1):102–114, February 2002.

[41] R. Nagarajan, J. Kurose, and D. Towsley. Local allocation of end-to-end quality-of-service in high-speed networks. *In IFIP TC6 Task Group/WG6.4 International Workshop on Performance of Communication Systems*, pages 99–118, January 1993.

[42] F. Kelly, A. Maulloo, and D. Tan. Rate control in communication networks: shadow prices, proportional fairness and stability. *Journal of the Operational Research Society*, 49, 1998.

[43] S. McCanne, V. Jacobson, and M. Vetterli. Receiver-driven layered multicast. *ACM SIGCOMM*, 26,4:117–130, August 1996.

[44] K. Kar, S. Sarkar, and L. Tassiulas. Optimization based rate control for multirate multicast sessions. *IEEE INFOCOM*, pages 123–132, 2001.

[45] K. Kar, S. Sarkar, and L. Tassiulas. A scalable low overhead rate control algorithm for multirate multicast sessions. *IEEE Journal of Selected areas in Communications, special issue in Network Support for Multicast Communications*, 20:1541–1557, October 2002.

[46] T. Nandagopal, T. Kim, X. Gao, and V. Bharghavan. Achieving MAC layer fairness in wireless packet networks. *In proceedings of the 6th annual international conference on Mobile computing and networking (MobiCom), Boston, MA, USA*, pages 87–98, 2000.

[47] L. Tassiulas and S. Sarkar. Maxmin fair scheduling in wireless networks. *In proceedings of 1EEE INFOCOM*, pages 763–772, 2002.

[48] H. Luo, S. Lu, and V. Bharghavan. A new model for packet scheduling in multihop wireless networks. *MobiCom, Boston, Massachusetts*, pages 76–86, 2000.

[49] C. Curescu and S. Nadjm-Tehrani. Price/utility-based optimized resource allocation in wireless ad hoc networks. *In proceedings of Second Annual IEEE Communications Society Conference on Sensor and Ad Hoc Communications and Networks (SECON)*, pages 85–95, September 2005.

[50] T. Salonidis and L. Tassiulas. Distributed dynamic scheduling for end-to-end rate guarantees in wireless ad hoc networks. *In proceedings of the 6th ACM international symposium on Mobile ad hoc networking and computing, Urbana-Champaign, IL, USA*, pages 145–156, 2005.

[51] S. Sarkar and L. Tassiulas. End-to-end bandwidth guarantees through fair local spectrum share in wireless ad-hoc networks. *In proceedings of IEEE Control and Decision Conference (CDC), Maui, HI, USA*, December 2003.

[52] L. Buttyan and J. P. Hubaux. Stimulating cooperation in self-organizing mobile ad hoc networks. *ACM/Kluwer Mobile Networks and Applications (MONET)*, 8(5), October 2003.

[53] Y. Qiu and P. Marbach. Bandwidth allocation in ad-hoc networks: A price-based approach. *In proceedings of IEEE INFOCOM*, 2003.

[54] J. Tang, G. Xue, C. Chandler, and W. Zhang. Link scheduling with power control for throughput enhancement in multihop wireless networks. *In proceedings of the second*

International Conference on Quality of Service in Heterogeneous Wired/Wireless Networks (QSHINE), Lake Buena Vista, FL, USA, August 2005.

[55] P. Djukic. Optimum resource allocation in multipath ad hoc networks. *M.Sc. Thesis University of Toronto*, 2003.

[56] T. M. Heikkinen. On congestion pricing in a wireless network. *Wireless Networks, Kluwer*, 8(4):347–354, july 2002.

[57] R. Liao, R. Wouhaybi, and A. Campbell. Incentive engineering in wireless lan based access networks. *In proceedings of 10th IEEE International Conference on Network Protocols (ICNP)*, November 2002.

[58] D. Julian, M. Chiang, D. ONeill, and S. Boyd. QoS and fairness constrained convex optimization of resource allocation for wireless cellular and ad hoc networks. *In proceedings of IEEE INFOCOM*, pages 477–486, June 2002.

[59] Y. Shavitt D. Raz. Optimal partition of QoS requirements with discrete cost functions. *IEEE Journal on Selected Areas in Communications*, 8, 2000.

[60] Y. Bejerano, Y. Breitbart, A. Orda, R. Rastogi, and A. Sprintson. Algorithms for computing QoS paths with restoration. *IEEE/ACM Transactions on Networking (TON)*, 13:648–661, June 2005.

[61] G. N. Rouskas and L. E. Jackson. Optimal granularity of MPLS tunnels. *In proceedings of the 18th International Teletraffic Congress (ITC-18), Berlin, Germany,* pages 1–10, September 2003.

[62] T. Jiang, M. H. Ammar, and E. W. Zegura. On the use of destination set grouping to improve inter-receiver fairness for multicast ABR sessions. *In proceedings of IEEE INFOCOM*, 2000.

[63] J. Turner. Terabit burst switching,. *Journal of High Speed Networks*, 8:3–16, 1999.

[64] R. Nagarajan, J. Kurose, and D. Towsley. Local allocation of end-to-end quality-of-service in high-speed networks. *In IFIP TC6 Task Group/WG6.4 International Workshop on Performance of Communication Systems*, pages 99–118, January 1993.

[65] J.F. Kurose and D. Towsley. Approximation techniques for computing packet loss in finite-buffered voice multiplexers. *IEEE Journal on Selected Areas in Communications*, 9(5), April 1991.

[66] S. Battiato, D. Cantone, D. Catalano, G. Cincotti, and M. Hofri. An efficient algorithm for the approximate median selection problem. *In proceedings of the Fourth Italian Conference, CIAC, Rome*, 1767:226–238, 2000.

[67] F. Gebali. *Computer Communication Networks: Analysis and Design*. Northstar Digital Design, 2005.

[68] R. Nagarajan, J.F. Kurose, and D. Towsley. Approximation techniques for comput-ing packet loss in finite-buffered voice multiplexers. *IEEE Journal on Selected Areas in Communications*, 9(5), April 1991.

[69] M. R. Garey and D. S. Johnson. *Computer & Intractability: A Guide to the Theory of NP-Completeness*. W H Freeman, November 1979.

[70] P. Gupta and P. Kumar. The capacity of wireless networks. *IEEE Transactions on Information Theory*, 46:388–404, March 2000.

[71] H. Gossain, N. Nandiraju, K. Anand, and D. P. Agrawal. Supporting MAC layer multicast in IEEE 802.11 based manets: Issues and solutions. *In proceedings of 29th Annual IEEE International Conference on Local Computer Networks (LCN)*, pages 172–179, 2004.

[72] P. Chaporkar, A. Bhat, and S. Sarkar. An adaptive strategy for maximizing through-put in MAC layer wireless multicast. *MobiHoc*, pages 256–267, 2004.

[73] D. Bertsekas. *Nonlinear Programming*. Athena Scientific, 1999.

[74] D. Bertsekas and J. Tsitsiklis. *Parallel and Distributed Computation*. Prentice-Hall, 1989.

[75] K. Nahrstedt S. H. Shah, K. Chen. Dynamic bandwidth management for single-hop ad hoc wireless networks. *ACM/Kluwer Mobile Networks and Applications (MONET)*, 10, 2005.

[76] E. M. Royer and C. E. Perkins. Multicast operation of the ad-hoc on-demand distance vector routing protocol. *ACM/IEEE International Conference on Mobile Computing and Networking (MOBICOM)*, pages 207–218, 1999.

[77] C. Perkins and E. Royer. Ad-hoc on-demand distance vector routing,. *In proceedings of 2nd IEEE Workshop on Mobile Computing Systems and Applications (WMCSA 99)*, 1999.

[78] J. G. Augustson and J. Minker. An analysis of some graph theoretical cluster techniques,. In *Journal of the Association for Computing Machinery*, volume 17, pages 571–586, 1970.

[79] E. Amir, S. McCanne, and R. Katz. An active service framework and its application to real-time multimedia transcoding. *In proceedings of the ACM SIGCOMM '98, Vancouver, Canada*, pages 178–189, 1998.

[80] N. Z. Shor. *Minimization Methods for Non-differentiable Functions*. Springer-Verlag, 1985.

[81] E. Rosen and A. Viswanathan. RFC 3031: Multiprotocol Label Switching Architecture. *Internet Engineering Task Force (IETF)*, January 2001.

[82] W. Rudin. *Principles of Mathematical Analysis*. McGraw-Hill, 1976.

Appendix A

QoS-Based Partitioning and Resource Allocation

This appendix presents the proofs for the QoS-based classification and partitioning in Chapters 3 and 4.

A.1 Proofs for Chapter 3

A.1.1 Lemma 3.1

Proof. Proceed by contradiction, assuming that we have an optimal classification π_L^* that is un-ordered, and for all ordered classifications π_{Lo} we have $\psi(\pi_L^*) < \psi(\pi_{Lo}) \ \forall \pi_{Lo}$. In other words, π_L^* is such that for some pairs of groups G_a and G_b, we have $r_i \in G_a$ and $r_j \in G_b$, and $Q_i > Q_j$ while $x_a \leq x_b$. We will define a re-arrangement operator η on π_L^*, and form another classification π_L' (i.e. $\pi_L' = \eta(\pi_L^*)$) and prove that by applying η repeatedly until π_L' becomes ordered, we will ultimately have $\psi(\pi_L') \leq \psi(\pi_L^*)$. Notice that for predetermined service levels, x_a and x_b may be allowed to have values which are outside of the range of QoS requests in G_a and G_b (e.g. $x_a < min(Q_i : Q_i \ \forall r_i \in Ga)$), and operator η will still work in this case. Here is how operator η will work.

1. If $Q_i \geq Q_j \geq x_b \geq x_a$ or $Q_i \geq x_b \geq Q_j \geq x_a$, η forms π'_L as follows:

$$G'_a = G_a - (r_i)$$

$$G'_b = G_b \cup (r_i)$$

$$\Delta\psi = U(Q_i, x_b) - U(Q_i, x_a) \leq 0$$

2. If $x_b \geq x_a \geq Q_i \geq Q_j$ or $x_b \geq Q_i \geq x_a \geq Q_j$, η forms π'_L as follows:

$$G'_b = G_b - (r_j)$$

$$G'_a = G_a \cup (r_j)$$

$$\Delta\psi = U(Q_j, x_a) - U(Q_j, x_b) \leq 0$$

3. If $x_b \geq Q_i \geq Q_j \geq x_a$, η forms π'_L as follows:

$$G'_b = (G_b \cup [r_i]) - [r_j] \;\rightarrow\; \psi(G_b) = U(Q_i, x_b) - U(Q_j, x_b) \leq 0$$

$$G'_a = (G_a \cup [r_j]) - [r_i] \;\rightarrow\; \psi(G_a) = U(Q_j, x_a) - U(Q_i, x_a) \leq 0$$

$$\Delta\psi = \Delta\psi(G_a) + \Delta\psi(G_b) \leq 0$$

By applying operator η $\forall r_i \in G_a$ and $\forall r_j \in G_b$ with $Q_i > Q_j$ and $x_a \leq x_b, \forall a$, and $\forall b$, we will obtain an ordered classification with $\psi(\pi'_L) \leq \psi(\pi^*_L)$, which contradicts the original assumption that π^*_L is an optimal classification. Therefore, $\psi(\pi'_L) = \psi(\pi^*_L)$, and π'_L is an optimal and ordered classification. $\qquad\square$

A.1.2 Lemma 3.2

Proof. Assuming the set of all possible ordered classifications for N traffic streams and L service levels is $\Im_L^N$, we proceed by induction. Clearly, the case of $L=1$ is trivial. For the case of $L=2$, the number of possible solutions ($|\Im_2^N|$) is N-1 which is $O(N)$. Now assume that for $L=m$, $|\Im_m^N|$ is $O(N^{m-1})$, to get $|\Im_{m+1}^N|$ when $L=m+1$, we notice that the number of possible solutions is actually derived by the summation of $|\Im_m^i|$ in the case of $i=m,\ m+1,\ldots,N-1$. This is because for ordered classifications, we can think of the groups as integer numbers and we want to get the possible combinations of numbers such that the total sum of all the numbers is N. Notice also that for $N < m$, $|\Im_m^N| = 0$. Therefore, for $L=m+1$, the possible combinations are the summation of $|\Im_m^i|$ for $i=m$, given that the last number is $N-m$, plus $|\Im_m^i|$ for $i=m+1$, given the last number is $N-m-1$, and so on. So, $|\Im_{m+1}^N|$ can be written as follows:

$$|\Im_{m+1}^N| = \sum_{i=m}^{N-1} |\Im_m^i|$$

It is not hard to prove that the summation part of this equation is $O(N)$ (e.g. $|\Im_4^N| = \sum_{i=3}^{N-1} (i-2) \times (i-1)/2 = 11/6 \times N - N^2 + 1/6 \times N^3 - 1 = O(N^3)$). So, the result $|\Im_L^N| = O(N^{L-1})$ follows. $\qquad\square$

A.1.3 Lemma 3.3

Proof. Since $a_{ij} = C_{ij} * U(Q_i/\omega_j, \gamma)$ is non-increasing in the interval $\gamma\omega_j \in [0, Q_i]$, and non-decreasing in the interval $\gamma\omega_j \in [Q_i, \infty[$, then $b_{ij} = U(Q_i/\omega_j, \gamma)$, is non-increasing for the values of $\gamma \in [0, Q_i/\omega_j]$, and non-decreasing for the values of $\gamma \in [Q_i/\omega_j, \infty[$. Without

loss of generality, assume the values of $d_{ij} = Q_i/\omega_j \ \forall i, \forall j$ are sorted in a non-decreasing

order. The summation for any possible subset of i, j of $\psi(\pi_L) = \sum_{i \in I} \sum_{j \in J} C_{ij} * U(Q_i/\omega_j, \gamma)$

is non-increasing in the interval of $\gamma \in [0, \min(Q_i/\omega_j)]$, and it is non-decreasing in the

interval $\gamma \in [\max(Q_i/\omega_j), \infty[, \forall i \in I \ \forall j \in J$. So, in order to minimize the value of $\psi(\pi_L)$,

γ must be selected in the interval $[\min(Q_i/\omega_j), \ \max(Q_i/\omega_j)] \ i = 1, ..., N \ j = 1,...,L.$ □

A.1.4   Theorem 3.1

Proof. Again, without loss of generality, we assume the values $d_{ij} = Q_i/\omega_j \ \forall i, \forall j$, are

sorted in a non-decreasing order. From the proof of Lemma 3.3, we know that the

summation $\Psi = \sum_{i \in I} \sum_{j \in J} a_{ij}$ for any subset of i, j is non-increasing for the values of $\gamma \in$

$[0, \min(d_{ij})]$, and non-decreasing for the values of $\gamma \in [\max(d_{ij}), \infty[\ , \forall i \in I \ \forall j \in J.$

Now, between any two subsequent values $d_{ij} = d_a$, and $d_{ij} = d_b$ such that $d_a \leq d_b$, the

summation Ψ for any subset of i, j is the summation of multiple concave functions. From

the properties of concave functions, the summation Ψ is a concave function in the range

$[d_a, d_b]$, which means that for any value $\gamma \in [d_a, d_b]$, we have $\Psi(\alpha \ d_a + (1 - \alpha)d_b) \ \geq$

$\alpha \ \Psi(d_a) + (1 - \alpha) \ \Psi(d_b) \ \ 0 \leq \alpha \leq 1.$ Therefore, to look for the minimum value of Ψ

within this range $\gamma \in [d_a, d_b]$, we only need to check at d_a, and d_b. We repeat the same

procedure for all successive values of $d_{ij} \ \forall i, \forall j$. So, the minimum value of Ψ for any

subset of i, j lies in the value set of $\gamma \in \{d_{ij}, i = 1, .., N \ j = 1, .., L\}$.   □

A.1.5 Theorem 3.2

Proof. Proceed by contradiction, assume that $\pi_L^*(\gamma')$ is the optimal classification for γ' such that $\psi(\pi_L^*(\gamma'), \gamma') \leq \psi(\pi_L^*(\gamma), \gamma) \,\forall\gamma$ and $\gamma' \notin \{d_{ij} = Q_i/\omega_j, \, i = 1, ..., N \quad j = 1, ..., L$. We have 3 scenarios for γ' as follows:

1. $\gamma' < \min(d_{ij})$

   We know from Theorem 3.1 that for the same classification $\pi_L^*(\gamma')$, we have

   $$\psi(\pi_L^*(\gamma'), \min(d_{ij})) \leq \psi(\pi_L^*(\gamma'), \gamma')$$

   We also know that considering the optimal classification for $\gamma = \min(d_{ij})$, we have

   $$\psi(\pi_L^*(\min(d_{ij})), \, \min(d_{ij})) \leq \psi(\pi_L^*(\gamma'), \, \min(d_{ij}))$$

   Therefore, $\psi(\pi_L^*(\min(d_{ij})), \min(d_{ij})) \leq \psi(\pi_L^*(\gamma'), \gamma')$, and this conflicts with the original assumption.

2. $d_a < \gamma' < d_b, \quad d_a > \min(d_{ij}) \quad , \quad d_b < \max(d_{ij})$

   Again, from Theorem 3.1, $\psi(\pi_L^*(\gamma'), \gamma')$ is concave in the interval $d_a < \gamma' < d_b$. So, for the same classification $\pi_L^*(\gamma')$, we have $\psi(\pi_L^*(\gamma'), d_a) \leq \psi(\pi_L^*(\gamma'), \gamma')$ and $\psi(\pi_L^*(\gamma'), d_b) \leq \psi(\pi_L^*(\gamma'), \gamma')$. We also know that $\psi(\pi_L^*(d_a), d_a) \leq \psi(\pi_L^*(\gamma'), d_a)$ and $\psi(\pi_L^*(d_b), d_b) \leq \psi(\pi_L^*(\gamma'), d_b)$. Therefore, $\psi(\pi_L^*(d_a), d_a) \leq \psi(\pi_L^*(\gamma'), \gamma')$ and $\psi(\pi_L^*(d_b), d_b) \leq \psi(\pi_L^*(\gamma'), \gamma')$ which conflicts with the assumption.

3. $\gamma' > \max(d_{ij})$

Similarly we can prove that $\psi(\pi_L^*(\max(d_{ij})),\ \max(d_{ij}))\ \leq\ \psi(\pi_L^*(\gamma'),\ \gamma')$ which

conflicts with the assumption.

Therefore, for all $\gamma' \notin \{d_{ij} = Q_i/\omega_j,\ i = 1, ..., N\ \ , j = 1, ..., L\}$, we cannot have

$\psi(\pi_L^*(\gamma'),\ \gamma') \leq \psi(\pi_L^*(\gamma),\ \gamma)\ \forall \gamma$. Then the values of

$\gamma^* \in \{d_{ij} = Q_i/\omega_j,\ i = 1, ..., N\ \ j = 1, ..., L\}$ constitute the optimal possible values

such that $\psi(\pi_L^*(\gamma^*),\ \gamma^*) \leq \psi(\pi_L^*(\gamma),\ \gamma)\ \ \forall \gamma$.   □

A.2   Proofs for Chapter 4

A.2.1   Lemma 4.1

Proof. To prove intractability, we will use the concept of restriction [69]. First, we can

easily convert the OQP minimization problem to a decision problem by assuming that we

will try to find whether there is a partition such that $\psi(P_L) = k$. Later we can decrease

k as low as possible to find the minimum value. Second, for $U(Q_i, s_l)$ defined by (4.7),

and for a specific partition P_L, we have for each group $G_l \in P_L$ of size $|G_l| = M$, we have

$M+1$ possible values for $\sum_{\forall Q_i \in G_l} U(Q_i, s_l) = \sum_{\forall Q_i \geq s_l} b_i + \sum_{\forall Q_i < s_l} a_i$ depending on which range

we select the value of s_l. Now, we are ready to do the transformation. By checking the

definition for the SUBSET SUM problem, we can map the set of sizes A to the possible

values of summation on each group such that $s(a) = \sum_{\forall Q_i \in G_l} U(Q_i, s_l)$. Remember for each

group in OQP we have $|G_l|$ possible numbers, which means we have $L*(|G_l|+1)$ possible

values for $s(a) \in A$ to cover all the L groups in OQP. Now, it is easy to see that finding the L subset $(A' \subset A)$ of numbers such that $\sum_{a \in A'} s(a) = B$ where B is positive integer is equivalent to finding the service levels such that $\psi(P_L) = k$, hence the result follows. $\square$

A.2.2 Theorem 4.1

Proof. Without loss of generality, we assume that the values of QoS are sorted such that $Q_1 \leq Q_2 \leq ... \leq Q_N$ and we prove by contradiction as follows. Assume that $P_L^* = \{G_1^*, G_2^*, ..., G_L^*\}$ is the optimal partition with a corresponding optimal service levels $S_L^* = \{s_1^*, s_2^*, ..., s_L^*\}$ such that S_L^* has some service levels $s_l^* \notin \{Q_i : i \in G_l\} \; \forall s_l^* \in S' \subset S_L$. In this case, we have the following 3 scenarios for each $s_l^* \in S'$:

1. $s_l^* < \min\{Q_i : i \in G_l\} = Q_l^{\min}$

   We know from the utility property that the summation $\sum_{\forall Q_i \in G_l} U(Q_i, s) = U(G_l \equiv \{Q_i, ..., Q_j\}, s)$ as a function of s will be non-increasing in the range $s \in [0, Q_l^{\min}]$ and non-decreasing in the range $s \in [Q_l^{\max}, \infty[$. This means $U(G_l, Q_l^{\min}) \leq U(G_l, s_l^*)$ and we can replace s_l^* by $Q_l^{\min}$ in S_L^*.

2. $s_l^* > \max\{Q_i : i \in G_l\} = Q_l^{\max}$

   We know from the utility property that the summation $U(G_l, s)$ will be non-increasing in the range $s \in [0, Q_l^{\min}]$ and non-decreasing in the range $s \in [Q_l^{\max}, \infty[$. This means $U(G_l, Q_l^{\max}) \leq U(G_l, s_l^*)$ and we can replace s_l^* by $Q_l^{\max}$.

3. $Q_m < s < Q_{m+1}$ $\quad m = 1, ..., N-1$

We notice for these ranges $U(G_l, s)$ is a summation of multiple concave functions. From the properties of concave functions, this summation is also concave, which means $U(G_l, Q_m) \leq U(G_l, s_l^*)$ and/or $U(G_l, Q_{m+1}) \leq U(G_l, s_l^*)$. In this case, we can replace s_l^* by either Q_m or Q_{m+1} (based on which one gives lower value for $U(G_l, s)$) in S_L^*.

Therefore, in the previous 3 scenarios, we can replace service levels in S' by values $s_l^* \in \{Q_i : i \in G_l\} \ \forall s_l^* \in S' \subset S_L$ and get lower value for the objective function which means that the replacement S_L^* is also optimal. $\qquad\qquad\square$

Appendix B

Resource Allocation for Wireless

Multicast: Convergence Analysis

This appendix presents the convergence analysis for our algorithm ORAWM for both synchronous and asynchronous environments.

B.1   Proof of Theorem 5.2

The proof follows the same way as Theorem 1 of [10]. First we prove the following lemma:

Lemma B.1. : *If u and v are any two feasible prices, i.e. $u, v \geq 0$, then based on Assumptions 1 and 2, ∇D satisfies the Lipscitch condition*

$$\|\nabla D(u) - \nabla D(v)\|_2 \leq \bar{Y}\bar{Z}/\bar{\gamma} \|u - v\|_2$$

Proof. from Equation (5.12), we have $\nabla D = C - \Gamma x$. Let $\frac{\partial x}{\partial p}(p)$ denote the $|M| \times |Q|$ matrix whose (m, q) element $\frac{\partial x_m}{\partial p_q}(p)$ is

$$\frac{\partial x_m}{\partial p_q}(p) = \begin{cases} \frac{\Gamma_{qm}}{U_m''(x_m(p))} & \text{if } U_m'(W_m) \leq p_q \leq U_m'(w_m) \\ \\ 0 & Otherwise \end{cases}$$

If we define $\beta_m(p)$ as follows:

$$\beta_m(p) = \begin{cases} -\frac{1}{U''_m(x_m(p))} & \text{if } U'_m(W_m) \leq p_q \leq U'_m(w_m) \\ \\ 0 & Otherwise \end{cases}$$

then $\frac{\partial x_m}{\partial p_q}(p)$ in matrix form can be written as

$$\left[\frac{\partial x_m}{\partial p_q}(p)\right] = -B(p) \; \Gamma^T$$

where $B(p) = Diag(\beta_m(p); m \in M)$ is the diagonal matrix with diagonal elements $\beta_m(p)$.

Hence,

$$\nabla^2 D = -\Gamma \left[\frac{\partial x_m}{\partial p_q}(p)\right] = \Gamma \; B(p) \; \Gamma^T \tag{B.1}$$

Now from Proposition A.25(e) in [74] and knowing that $\nabla^2 D = \Gamma \; B(p) \; \Gamma^T$ is symmetric

(i.e. $\left\|\Gamma \; B(p) \; \Gamma^T\right\|_1 = \left\|\Gamma \; B(p) \; \Gamma^T\right\|_\infty$), then we have

$$\begin{aligned} \left\|\Gamma \; B(p) \; \Gamma^T\right\|_2 \quad &\leq \left\|\Gamma \; B(p) \; \Gamma^T\right\|_\infty \\ &= \max_q \sum_{q'} \left[\Gamma \; B(p) \; \Gamma^T\right]_{qq'} \\ &= \max_q \sum_{q'} \sum_{m} \beta_m(p) \; \Gamma_{qm} \; \Gamma_{q'm} \\ &= \max_q \sum_{m} \beta_m(p) \; \Gamma_{qm} \sum_{q'} \Gamma_{q'm} \\ &\leq \bar{Y}\bar{Z}/\bar{\gamma} \tag{B.2} \end{aligned}$$

From Theorem 9.19 in [82] we have for equation (B.2)

$$\left\|\nabla D(u) - \nabla D(v)\right\|_2 \leq \bar{Y}\bar{Z}/\bar{\gamma} \left\|u - v\right\|_2$$

and hence the result follows. $\square$

Proof. (Theorem 5.2) from lemma B.1, the dual objective function D is lower bounded and ∇D is Lipschitz. Then, limit point p^* of the sequence $\{p(t)\}$ generated by the gradient projection algorithm in Figure 5.4 for the dual problem is dual optimal provided that $0 < \alpha < 2\bar{\gamma}/\bar{Y}\bar{Z}$ (see Proposition 3.4 in [74]).

let $\{p(t)\}$ be a subsequence converging to p^*. Since $U'_m(x_m)$ is defined on a compact interval $[w_m, W_m]$, it is continuous and one-to-one (because of the strict concavity of $U_m(x_m)$). Thus, its inverse is continuous (see Theorem 4.17 in [82]) and hence from Equation (5.15), $x(p)$ is continuous. Therefore, $\lim_{t\to\infty} x(t) = x(p^*)$ and that proves the result of Theorem 5.2. □

B.2 Proof of Theorem 5.3

The proof for this theorem uses a similar approach like the one in Theorem 3 [6]. However, the difference in our case is that we consider asynchrony amongst the terminal nodes in calculating the accumulated price, which is relevant to the multicast case.

We define a vector $\pi(t) = p(t+1) - p(t)$ which measures the successive price change with time. First, we prove that the error in rate calculation of group m is bounded by the successive price change π:

Lemma B.2. : *The estimated price for group m at time t is defined as follows:*

$$\hat{\lambda}_m(t) = \sum_{i:f_{mi} \in F_m} \sum_{q:f_{mi} \in V_c^q} \sum_{t''=t-2B}^{t} \epsilon_{qt''} p_q(t'')$$

where $\epsilon_{qt''}$ is defined by a function $\epsilon(t) = (\epsilon_{qt''}(t), q \in Q, t'' \in [t-2B, t])$ as follows:

$$\epsilon_{qt''}(t) = \begin{cases} b_q^i(t'', t) \cdot \sum_{t'=t-B}^{\min\{t''+B,t\}} b_m^i(t', t) & \text{,if } f_{mi} \in V_c^q \\ \\ 0 & \text{Otherwise} \end{cases}$$

and $\bar{p}_q(t) = (p_q(t''), t'' \in [t-2B, t])$ is the sequence of clique q's price at time instances $t - 2B, t - 2B + 1, \cdots, t$.

Proof. From Equation (5.16) we can write the accumulated price for one branch from group source s_m to one of the terminal nodes $i \in \Im_m$ as follows:

$$\lambda_m(i, t+1) = \sum_{j \in s_m \Rightarrow i} \left(\prod_{j \in s_m \Rightarrow i} 1/k_{mj} \right) \hat{\lambda}_{mj}(t) \tag{B.3}$$

The estimated value at time t is then defined by

$$\hat{\lambda}_m(i,t) = \sum_{t'=t-B}^{t} b_m^i(t',t) \sum_{j\in s_m \Rightarrow i} \left(\prod_{j\in s_m \Rightarrow i} 1/k_{mj} \right) \hat{\lambda}_{mj}(t)$$

$$= \sum_{j\in s_m \Rightarrow i} \left(\prod_{j\in s_m \Rightarrow i} 1/k_{mj} \right) \sum_{t''=t-2B}^{t} \epsilon_{qt''}(t) p_q(t'')$$

because

$$\sum_{t'=t-B}^{t} b_m^i(t') \cdot \sum_{t''=t'-B}^{t'} b_q^i(t'') p_q(t'') = \sum_{t''=t-2B}^{t} \left(b_q^i(t'') \cdot \sum_{t'=t-B}^{\min\{t''+B,t\}} b_m^i(t') \right) p_q(t'')$$

From Theorem 5.1, the group price is then the summation of all the accumulated prices on all the branches and can be written as follows (see proof of Theorem 5.1):

$$\hat{\lambda}_m(t) = \sum_{i:f_{mi}\in F_m} \sum_{q:f_{mi}\in V_c^q} \sum_{t''=t-2B}^{t} \epsilon_{qt''} p_q(t'')$$

which proves the lemma. □

Lemma B.3. :

a) For all t

$$\left| U_m'^{-1}(\hat{\lambda}_m(t)) - U_m'^{-1}(\lambda_m(t)) \right| \le 1/\gamma_m \sum_{t''=t-2B}^{t-1} \sum_{q\in Q} \left| \pi_q(t'') \right| \Gamma_{qm}$$

b) For all t

$$\left| U_m'^{-1}(\lambda_m(t)) - U_m'^{-1}(\lambda_m(\tau)) \right| \le 1/\gamma_m \sum_{t'=\tau}^{t-1} \sum_{q\in Q} \left| \pi_q(t') \right| \Gamma_{qm}$$

Proof. First, we denote $x_m(\epsilon; \bar{p}(t))$ as

$$x_m(\epsilon; \bar{p}(t)) = U_m'^{-1} \left(\sum_{i:f_{mi}\in F_m} \sum_{q:f_{mi}\in V_c^q} \sum_{t''=t-2B}^{t} \epsilon_{qt''} p_q(t'') \right) \tag{B.4}$$

where $\bar{p}(t)$ is the vector of the sequence $\bar{p}_q(t)$ $\forall q \in Q$. Hence

$$\frac{\partial x_m(\epsilon; \bar{p}(t))}{\partial \epsilon_{qt''}} = \frac{\Gamma_{qm} p_q(t'')}{U''_m(x_m(\epsilon; \bar{p}(t)))}$$

and by Assumption 2, we have that

$$0 \leq \left| \frac{\partial x_m(\epsilon; \bar{p}(t))}{\partial \epsilon_{qt''}} \right| \leq (1/\gamma_m) \Gamma_{qm} p_q(t'') \tag{B.5}$$

we also denote $1(t) = (1_{qt''}(t)), q \in Q, t'' \in [t - 2B, t])$ as

$$1_{qt''}(t) = \begin{cases} 1 & , \text{if } F_m \cap V_c^q \neq \emptyset, t'' = t \\ \\ 0 & , Otherwise \end{cases}$$

Therefore, $U'^{-1}_m(\hat{\lambda}_m(t)) = x_m(\epsilon(t); \bar{p}(t))$ and $U'^{-1}_m(\lambda_m(t)) = x_m(1(t); \bar{p}(t))$, where $x_m(., \bar{p}(t))$ is defined in Equation (B.3). By the mean value theorem (see Proposition A.22 in [73]), we have for some $\tilde{\epsilon}$,

$$\left| U'^{-1}_m(\hat{\lambda}_m(t)) - U'^{-1}_m(\lambda_m(t)) \right| = \left| \sum_{q \in Q, t'' \in [t-2B,t]} \frac{\partial x_m(\tilde{\epsilon}; \bar{p}(t))}{\partial \epsilon_{qt''}} (1_{qt''}(t) - \epsilon_{qt''}) \right|$$

$$\leq 1/\gamma_m \left| \sum_{q \in Q, t'' \in [t-2B,t]} \Gamma_{qm} \bar{p}_{qt''}(t)(1_{qt''}(t) - \epsilon_{qt''}) \right|$$

$$\leq 1/\gamma_m \sum_{q \in Q} \Gamma_{qm} \max_{t-2B \leq t'' \leq t} \left| p_q(t) - p_q(t'') \right|$$

$$\leq 1/\gamma_m \sum_{t''=t-2B}^{t-1} \sum_{q \in Q} \Gamma_{qm} \left| \pi_q(t'') \right|$$

Similarly, also by the mean value theorem, we have that

$$\left| U'^{-1}_m(\lambda_m(t)) - U'^{-1}_m(\lambda_m(\tau)) \right| \leq 1/\gamma_m \sum_{q \in Q} \Gamma_{qm} |p_q(t) - p_q(\tau)|$$

$$\leq 1/\gamma_m \sum_{t'=\tau}^{t-1} \sum_{q \in Q} \Gamma_{qm} \left| \pi_q(t') \right|$$

which proves lemma B.3. □

The next step is to formulate error in estimating the gradient for the dual objective function in the same way. To do that, we denote the vector $\xi = (\xi_q(t), q \in Q)$ to be the gradient estimation used in our asynchronous algorithm.

Lemma B.4. *There exists a constant $K_1 > 0$ such that*

$$\|\nabla D(p(t)) - \xi(t)\| \leq K_1 \sum_{t'=t-2B}^{t-1} \left\| \pi(t') \right\| \tag{B.6}$$

Proof. First from equation (5.12) and the clique procedure in Figure 5.5 we have that

$$[\nabla D(p(t)) - \xi(t)]_q = \sum_{m:(F_m \cap V_c^q) \neq \emptyset} \Gamma_{qm} \left(\sum_{t'=t-B} b_i^q(t',t) x_m(t') - \bar{x}_m(t) \right)$$

Where $\bar{x}_m(t)$ is the rate of group m if the group source knows the exact group price $\lambda_m(t)$. By Proposition A.2 in [74] (i.e. $\|.\|_2 \leq K_1' \|.\|_\infty$), we have that

$$
\begin{aligned}
\|\nabla D(p(t)) - \xi(t)\| \quad &\leq K_1' \max_{q \in Q} \sum_{m:F_m \cap V_c^q \neq \emptyset} \Gamma_{qm} \left| \sum_{t'=t-B} b_i^q(t',t) x_m(t') - \bar{x}_m(t) \right| \\
&\leq K_1' \max_{q \in Q} \sum_{m:F_m \cap V_c^q \neq \emptyset} \Gamma_{qm} \max_{t-B \leq t' < t} \left| x_m(t') - \bar{x}_m(t) \right| \\
&\leq K_1' \max_{q \in Q} \sum_{m:F_m \cap V_c^q \neq \emptyset} \Gamma_{qm} \max_{t-B \leq t' < t} \left| U_m'^{-1}(\hat{\lambda}_m(t')) - U_m'^{-1}(\lambda_m(t)) \right|
\end{aligned}
$$

Applying Lemma B.3, we have

$$\|\nabla D(p(t)) - \xi(t)\|$$

$$\leq K_1' \max_{q \in Q} \sum_{m:F_m \cap V_c^q \neq \emptyset} \Gamma_{qm}$$

$$\left(\max_{t-B \leq t' < t} \left| U_m'^{-1}(\hat{\lambda}_m(t)) - U_m'^{-1}(\lambda_m(t')) \right| + \left| U_m'^{-1}(\lambda_m(t')) - U_m'^{-1}(\hat{\lambda}_m(t')) \right| \right)$$

$$\leq K_1' \max_{q \in Q} \sum_{m:F_m \cap V_c^q \neq \emptyset} \Gamma_{qm}$$

$$\left(\max_{t-B \leq t' < t} 1/\gamma_m \left\{ \sum_{\tau=t''}^{t-1} \sum_{q' \in Q} \Gamma_{q'm} \left| \pi_{q'}(\tau) \right| + \sum_{\tau=t''-2B}^{t''-1} \sum_{q' \in Q} \Gamma_{q'm} \left| \pi_{q'}(\tau) \right| \right\} \right)$$

$$= K_1' \max_{q \in Q} \sum_{m:F_m \cap V_c^q \neq \emptyset} \Gamma_{qm} \max_{t-B \leq t' < t} 1/\gamma_m \sum_{\tau=t''-2B}^{t-1} \sum_{q' \in Q} \Gamma_{q'm} \left| \pi_{q'}(\tau) \right|$$

$$\leq K_1' \bar{Z}\bar{Y}/\bar{\gamma} \sum_{\tau=t-3B}^{t-1} \|\pi(\tau)\|_1$$

where the last inequality follows from Equation (B.2). Note that if we prove the convergence with respect to $\|\pi(t)\|_1$, then the convergence happens with respect to all other norms (see Proposition A.9 in [73]). Hence, if we let $K_1 = K_1' \bar{Z}\bar{Y}/\bar{\gamma}$, then this proves Lemma B.4. $\qquad\square$

Now that we formulated the gradient in terms of $\|\pi(t)\|$, we now show that this norm sequence converges to zero in the following lemma.

Lemma B.5. *Provided α is sufficiently small we have $\|\pi(t)\| \to 0$ as $t \to \infty$.*

Proof. First, by Lemma 5.1 in [74] (Section 7.5.4), we have for all t, $\xi^T(t)\pi(t) \leq -(1/\alpha) \|\pi(t)\|^2$. By Proposition A.32 in [74] and Lemma B.1, there exists K_2 such that

$$D(p(t+1)) \leq D(p(t)) + \|\nabla D(p(t)) - \xi\| \cdot \|\pi(t)\| - \left(\tfrac{1}{\alpha} - K_2\right) \|\pi(t)\|^2$$

applying Lemma B.4 we have

$$D(p(t+1)) \leq D(p(t)) - \left(\frac{1}{\alpha} - K_2\right) \|\pi(t)\|^2 + K_1 \sum_{t'''=t-3B}^{t-1} \left\|\pi(t''')\right\| \cdot \|\pi(t)\|$$

Now it can be shown that

$$\sum_{t'''=t-3B}^{t-1} \left\|\pi(t''')\right\| \cdot \|\pi(t)\| \leq \frac{3}{2} B \|\pi(t)\|^2 + \frac{1}{2} \sum_{t'''=t-3B}^{t-1} \left\|\pi(t''')\right\|^2$$

Then we have

$$D(p(t+1)) \leq D(p(t)) - \left(\frac{1}{\alpha} - K_2 - (\frac{3}{2}B - \frac{1}{2})K_1\right) \|\pi(t)\|^2 + \frac{K_1}{2} \sum_{t'''=t-3B}^{t-1} \left\|\pi(t''')\right\|^2$$

Now, applying this inequality recursively to all $D(p(t)), \tau = t, t-1, \cdots, 1$, taking $\pi(t) = 0$

for $t < 0$, we have

$$D(p(t+1)) \leq D(p(0)) - \left(\frac{1}{\alpha} - K_2 - (3B+1)K_1\right) \sum_{\tau=0}^{t} \|\pi(\tau)\|^2$$

By choosing α sufficiently small such that $\left(\frac{1}{\alpha} - K_2 - (3B+1)K_1\right) > 0$ and since

$D(p(t))$ is bounded, then as $t \to \infty$, we must have $\sum_{t=0}^{\infty} \|\pi(t)\|^2 < \infty$, and hence $\|\pi(t)\| \to$

0 as $t \to \infty$. This proves Lemma B.5. $\qquad\square$

Proof. (Theorem 5.3) We first prove that the error due to asynchronism for calculating

the group price and rates all converge to zero. Using Lemma B.2, we have that

$$|\hat{\lambda}_m(t) - \lambda_m(t)| =$$

$$|\sum_{i\in\mathfrak{I}_m} \sum_{t'=t-B}^{t} b_m^i(t',t) \sum_{j\in s_m \Rightarrow i} (\prod_{j\in s_m \Rightarrow i} 1/k_{mj}) \sum_{q\in Q} \sum_{t''=t'-B}^{t'} b_q^i(t'')p_q(t'')$$

$$- \sum_{i\in\mathfrak{I}_m}\sum_{j\in s_m\Rightarrow i} (\prod_{j\in s_m\Rightarrow i} 1/k_{mj}) \sum_{q:f_{mi}\in V_c^q} p_q(t)| =$$

$$| \sum_{i:f_{mi}\in F_m} \sum_{q:f_{mi}\in V_c^q} \sum_{t'=t-B}^{t} b_m^i(t') \cdot \sum_{t''=t'-B}^{t'} b_q^i(t'')p_q(t'') - \sum_{i:f_{mi}\in F_m} \sum_{q:f_{mi}\in V_c^q} p_q(t)|$$

Similarly as we did in Lemma B.3, we have that

$$|\hat{\lambda}_m(t) - \lambda_m(t)| = \quad \leq \sum_{q \in Q} \Gamma_{qm} \max_{t-2B \leq t'' \leq t} |p_q(t) - p_q(t'')|$$

$$\leq \sum_{q \in Q} \Gamma_{qm} \sum_{t''=t-2B}^{t} \left\| \pi(t'') \right\|_1$$

which by Lemma B.5 converges to 0 as $t \to \infty$. Because $x_m(t)$ and $\bar{x}_m(t)$ are projections of U'^{-1}_m onto $I_m = [w_m, W_m]$ and projection is nonexpansive (see Proposition 2.1.3 in [73]), we have that

$$|x_m(t) - \bar{x}_m(t)| \quad \leq |U'^{-1}_m(\hat{\lambda}_m(t) - U'^{-1}_m(\lambda_m(t))|$$

$$\leq (1/\gamma_m) \sum_{q \in Q} \Gamma_{qm} \sum_{t''=t-2B}^{t} \left\| \pi(t'') \right\|_1$$

which also by Lemma B.5 converges to 0 as $t \to \infty$ for all $m \in M$.

We now show that every limit point of the sequence $\{p(t)\}$ generated by the asynchronous algorithm minimizes the dual problem. Let p^* be a limit point of $\{p(t)\}$. At least one exists, as it is constrained to lie in the compact set, provided α is sufficiently small. Moreover, since the interval between consecutive updates is bounded (Assumption 3), it follows that there exists a sequence of elements of T along which p converges. Let $\{t_k\}$ be a subsequence such that $\{p(t_k)\}$ converges to p^*. By Lemma B.4, we have

$$\lim_k \xi(t_k) = \lim_k \nabla D(p(t_k)) = \nabla D(p^*)$$

Hence

$$[p^* - \alpha \nabla D(p^*)]^+ - p^* = \lim_k [p(t_k) - \alpha \, \xi(t_k)]^+ - p(t_k) = \lim_k \pi(t_k) = 0$$

Then by the projection theorem (Propositions 2.1.3 in [73] and 3.3 in [74]), p^* minimizes D over $p \geq 0$. By duality $x^* = x(p^*)$ is the unique primal optimal rate. We now show that it is a limit point of $\{x(t)\}$ generated by asynchronous algorithm. Consider a subsequence $\{x(t_m)\}$ of $\{x(t_k)\}$ such that $\{x(t_m)\}$ converges. Since $\|x(t) - \bar{x}(t)\| \to 0$, then

$$\lim_k x(t_m) = \lim_k \bar{x}(t_m) = \lim_k x(p(t_m)) = x(p^*)$$

This completes the proof for Theorem 5.3 □

Wissenschaftlicher Buchverlag bietet

kostenfreie

Publikation

von

wissenschaftlichen Arbeiten

Diplomarbeiten, Magisterarbeiten, Master und Bachelor Theses
sowie Dissertationen, Habilitationen und wissenschaftliche Monographien

Sie verfügen über eine wissenschaftliche Abschlußarbeit zu aktuellen oder zeitlosen
Fragestellungen, die hohen inhaltlichen und formalen Ansprüchen genügt,
und haben **Interesse an einer honorarvergüteten Publikation**?

Dann senden Sie bitte erste Informationen über Ihre Arbeit per Email
an info@vdm-verlag.de. Unser Außenlektorat meldet sich umgehend bei Ihnen.

VDM Verlag Dr. Müller Aktiengesellschaft & Co. KG
Dudweiler Landstraße 125a
D - 66123 Saarbrücken

www.vdm-verlag.de

www.ingramcontent.com/pod-product-compliance
Lightning Source LLC
Chambersburg PA
CBHW071425050326
40689CB00010B/1986